THERMODYNAMICS
© Scott Post 2020

All rights reserved. no part of this book may be reproduced without permission of the author.

He aha te mea nui o te ao
What is the most important thing in the world?
He tangata, he tangata, he tangata
It is the people, it is the people, it is the people
Maori proverb

Table of Contents

INTRODUCTION	2
UNITS	4
CONSERVATION OF ENERGY	7
Energy Use in the US	10
Wind Energy	14
Energy Storage	16
Enthalpy	19
First Law of Thermodynamics	22
WASTE HEAT	23
Entropy	23
HEAT TRANSFER	26
THERMODYNAMIC PROPERTIES & PHASES OF MATTER	30
Gibbs Phase Rule	39
Ideal Gas Law	40
IDEAL GAS MIXTURES	44
SIMPLE DEVICES	48
Piston-Cylinder	48
Water pumps	52
Compressors	55
Turbines	59
Electric motors and batteries	60
MULTI-COMPONENT DEVICES	64
Thermal Power Plants	64
Steam Power Plants	64
Carnot Cycle	69
Endoreversible Carnot Cycle	71

GAS TURBINE	75
REFRIGERATORS	79
TANK FILLING AND EMPTYING	85

SIMPLIED FORMS OF ENERGY EQUATION — 93

PSYCHROMETRICS — 96

GENERAL TWO-PHASE MIXTURES — 101

COMBUSTION — 103

REFERENCES — 107

THERMODYNAMICS EQUATION SHEET — 109

USEFUL WEBSITES — 111

NOTE TO INSTRUCTORS — 113

APPENDICES: DATA TABLES — 114

USING TABULAR DATA TABLES	114
LINEAR INTERPOLATION	115
VAPOR DOMES	116
SUMMARY TABLES	119
PROPERTIES OF WATER	120
PROPERTIES OF AIR	124
CONVERSION FACTORS	127

MATLAB CODES — 128

Introduction

If you are reading this book, it is most likely because either you are struggling in your thermodynamics course and looking for help, or you don't want to spend close to $200 for a textbook and are looking for a cheaper alternative.

This book is designed to help you solve practical thermal engineering problems. I believe it is possible to do this without having a deep philosophical understanding of entropy (though I hope reading this book does help you understand entropy better). So if your instructor is giving you quizzes with questions asking you about irreversibility, the Clausius inequality, or the Kelvin-Planck statement of the second law, this book will not help. This book is designed only to help you solve (practical) problems, relating to important engineering devices, including internal combustion engines, thermal power plants, gas turbine engines, compressors, pumps, and turbines, refrigerators and heat pumps.

The key concepts are:
- Conservation of Energy
- Waste Heat
- Efficiency
- Ideal Processes
- Thermodynamic Properties

Key Resources in this book:
- Worked example problems, including some using property data available online rather than printed data tables
- Equation sheet – useful for exams
- List of useful websites for videos and more reading on topics of interest to you
- Property data tables for water and air
- Where to find data – online resources for finding thermodynamic properties

My advice to students – Ask yourselves: What is the most important thing? To believe that you are capable of understanding and solving thermodynamic problems. In fact, anyone is capable of doing this, IF they put in the time to up-skill and practice learning what is necessary to be a successful engineer. You will not be successful in thermodynamics (or anything else) if you do not put in the time. Of course, you can waste a lot of time doing the same thing over and over again if it is not productive. So if what you have been doing so far to be successful in your coursework is not successful, you need to try something else. You can get help from a classmate, an upperclassman who was taken the course before, a tutor, online resources (see list at the end of this book), or the library. There is no need suffer through a class you are not doing well in and continue suffering the whole semester. Also, you don't have to understand everything about thermo - just enough to work engineering problems you will find in real life.

What knowledge this book assumes you already have knowledge of:
- Ideal gas law
- Math – basic calculus, including the ability to calculate integrals and derivatives, balancing equations, non-linear equations
- Chemistry – elements, atoms, and molecules
- Physics – kinetic and potential energy

What is a **model** – It is a way to approximate or simplify something that happens in the real world down to a level where we can understand it and mathematically approach it with some type of analytical equation. As an example – when we work problems on the piston-cylinder motion in an internal combustion engine, we will *model* the compression stroke of the engine as an isentropic process of an ideal gas (don't worry, I will explain isentropic later). This allows us to use a simple formula to relate the pressure of the gas in the cylinder to the position of the piston at any point of its motion.

We start this book with a discussion of units – oftentimes you can solve a problem simply by matching the units you need with the data you have

Units

Every problem in this book is in metric units. I understand that if you want to work in engineering in the U.S. you will need to be able to work with the English (Imperial or non-metric) units. I have found in my experience the most sure-fire way to ensure I don't make a mistake when working in English units is to first convert all my variables into metric units, work my calculations, and then convert back to English units at the end.

It is possible your instructor will make you work through calculations entirely in English units. I cannot help with that. In this book all the energy and power calculations in metric units (as any sane person would).

Fundamental metric units: In the m-k-s unit system, most other units can be created by combining length (m), mass (kg), and time (s) in the needed ratios. For example, velocity (m/s) is length divided by time. Eventually we also need to add units for temperature (°C) and electric charge (coulombs or C) to cover thermal and electrical properties, respectively. Below is shown how the units for pressure, force, energy, and power are combinations of mass, length, and time:

Pressure

$$1 \, Pa = 1 \, \frac{N}{m^2} = 1 \frac{kg}{m \cdot s^2}$$

$$1 \, psi = 1 \, \frac{lbf}{in^2} = 144 \frac{lbf}{ft^2}$$

Force

$$1 \, N = 1 \, kg \frac{m}{s^2}$$
$$1 \, lbf = 32.2 \, lbm \frac{ft}{s^2}$$

Energy

$$1 \, J = 1 \, N \times m = 1 \, kg \frac{m^2}{s^2}$$

$$1\, Btu = 778\, ft \times lbf = 25{,}058\, lbm\frac{ft^2}{s^2}$$

Power

$$1\, W = 1\frac{J}{s} = 1\, kg\frac{m^2}{s^3}$$

$$1\, hp = 550\, \frac{lbf \times ft}{s} = 17{,}710\, lbm\frac{ft^2}{s^3}$$

In English units I use lbm for pound mass and lbf for pound force.

- 1 N = 1 kg*m/s² (remember F = m*a)
- 1 J = 1 N * m = 1 kg*m²/s² is the fundamental unit of energy
- 1 W = 1 J / s is the unit of power (energy rate or energy/time)

If you are forced to use English units, the following conversions will be helpful –

Conversion factors:

1 hp = 550 ft-lbf/s	4.45 N = 1 lb 1 m = 3.28 ft
1 lbf = 32.2 lbm-ft/s²	1 hp = 746 W
1 m/s = 2.24 mph	1 mile = 5280 ft
1 Gal = 231 in³ = 3.785 L	1000 L = 1 m³
2.54 cm = 1 in	1 kg = 2.2 lbm
1 atm = 14.7 psi = 101,325 Pa	1 bar = 100,000 Pa = 14.5 psi
1 ft-lbf = 1.357 N-m	1 mi = 5280 ft = 1.609 km = 1609 m
[°F] = 9/5 [°C] + 32	[°C] = ([°F] - 32)*5/9
1 mpg = 0.425 km/L	

When used as a unit of energy, the *British Thermal Unit* (BTU) is equivalent to: 1 BTU = 778 ft*lbf. Just to make things even more confusing with English units, a Btu sometimes also refers to cooling power, in Btu/hr. When used as a unit of cooling power, (BTU/hour), it can be related to mechanical power as: 1 hp = 2544 Btu/hr. It is important to know the difference between energy and power (the rate at which energy is used). Further conversion factors are included in the Appendix at the end of this book.

1 kW*hr is the amount of energy used when 1 kilowatt of power is used for an hour. Your electricity bill normally tells how many kilowatt-hours you used that month, and the electricity is priced in units of $/kW-hr. For example, in St. Louis, MO, USA the price as of May 2019 is US$0.097/kWh = 9.7¢/kWh, while in Christchurch, New Zealand, the price is NZ$0.25/kWh = US$0.16/kWh. (Sometimes people use *h* for hours and other times *hr*).

Note: 1 kW = 1 kilowatt = 1000 watts

So let's consider an example of whether it makes economic sense to buy an LED light bulb or an incandescent light bulb the next time one of your light bulbs burns out. We will compare a 4.5 W LED bulb that puts out the same amount of light (470 lumens) as a 40 W incandescent bulb. The LED cost NZ$8.95 and 40 W incandescent bulb cost NZ$0.86.

So if light is left on 2 hours per day in the evening every day, it is used for 365*2 = 730 hours over a year. Energy use for the incandescent bulb over one year is: 40 W * 730 hours = 29200 Watt-hours = 29.2 kW*hr. The cost of electricity to power that bulb in New Zealand is 29.2*0.16 = $4.67. Energy use for the LED bulb over a year is: 4.5 W * 730 hours = 3825 Watt-hours = 3.8 kW*hr. The cost of electricity is 3.8*0.16 = $0.61.

We can calculate the payoff period for the capital investment in the more expensive LED bulb by making the following equation and solving for X:

$$X \text{ years} * (\$4.67 - 0.61) = \$8.96 - \$0.86 = \$8.10$$

So it would take 2.0 years for the LED bulb to pay off.

Conservation of Energy

The *First Law of Thermodynamics* tell us:
"Energy cannot be created or destroyed, only transferred from one form to another."

In other words, there is no free energy. To get energy you have to take it from somewhere or convert it from another form. As an example, hydrogen is not an energy source, it is an energy carrier. There are no large stores hydrogen in the earth's crust waiting to be mined and exploited. In order to use hydrogen as the fuel in a combustion engine or an electro-chemical fuel cell, we must first create the hydrogen, most commonly by electrolysis from water. Whether we burn hydrogen as a fuel or use it in a fuel cell, we must add oxygen to release the chemical energy, and the resultant product of a hydrogen-oxygen reaction is water (H_2O). Thus we have ended up back where we started, with water. The amount of energy it takes to create hydrogen from water is exactly the same as the amount of energy released when we react hydrogen to create water, so there is zero net energy gain. In fact you may have already noticed in your experience, no process is 100% efficient – there are always some energy losses when we are moving energy around (this observation leads us to the *2nd Law of Thermodynamics*, covered in the next section).

Forms of energy
- Kinetic
- Potential (Gravitational)
- Electrical
- Thermal
- Chemical
- Nuclear

So let's talk about the different forms energy can take, using a common car (a Toyota Corolla seems to be a typical student vehicle) to illustrate. A car can contain the following forms of energy:

Type of energy	Example in automobile
Chemical energy	Gasoline in the fuel tank
Translational kinetic energy	Forward motion of the vehicle
Rotational kinetic energy	Spinning wheels
Gravitational potential energy	Change in elevation going up or down a hill
Electrical energy	Battery

The Specifications for car are:

Quantity	Value
Mass	1305 kg (2870 lbm)
Driving speed	26.8 m/s (60 mph)
Engine size	1.8 L, 4-cylinder
Rated engine power	98.5 kW (132 hp)
Fuel tank capacity	50.3 L (13.3 gal)
Drag area	0.606 m^2
Fuel economy	7.8 L/km (30 mpg)

Kinetic Energy

So to calculate the kinetic energy (KE), we need the velocity in units that are convenient to work with, namely m/s, since 1 Joule = 1 kg (m/s)^2. So taking a cruising speed of 60 miles per hour, we can convert that to m/s:

$$60 \text{ mph} = 96.5 \text{ kph} = 26.8 \text{ m/s}$$

And now we can calculate the kinetic energy in standard metric units:

$$KE = \tfrac{1}{2} mV^2 = 0.5(1305 \text{ kg})(26.8 \text{ m/s})^2 = 469234 \text{ kg m}^2/\text{s}^2 = 469{,}200 \text{ J} = 469.2 \text{ kJ}$$

This is the energy the brakes must dissipate to bring the car to a complete stop.

Chemical energy

How much energy is stored in the fuel tank? = The fuel tank size = 13.3 gal = 50.3 L, so the mass of fuel in the tank is

$$m = \rho V = (50.3 \text{ L})(0.75 \text{ kg/L}) = 37.7 \text{ kg}$$

Chemical energy of fuels (see combustion section at the end of this book) is reported as *heating values*, in units of kJ/kg. For most petroleum-derived hydrocarbon fuels, such as gasoline/petrol, heating values are in the range of 40,000-45,000 kJ/kg. Using a value of 44,000 kJ/kg for gasoline, we can calculate the chemical energy contained in a full fuel tank:

$$(50.3 \text{ L})(0.75 \text{ kg/L})(44{,}000 \text{ kJ/kg}) = 1{,}659{,}000 \text{ kJ} = 1.659 \times 10^9 \text{ J}$$

Potential energy (PE)

If the car is on top of a hill 100 m (328 ft) above its surroundings, what is its gravitational potential energy?

$$PE = mgh = (1305 \text{ kg})(9.8 \text{ m/s}^2)(100 \text{ m}) = 1{,}278{,}900 \text{ J} = 1{,}279 \text{ kJ}$$

Electrical energy

A typical car battery might be rated for 60 amp-hours (Ah) at 12 V – what is the electrical energy it contains?

$$60 \text{ A} * 12 \text{ V} * 1 \text{ hr} = 720 \text{ W*h} * 3600 \text{s/h} = 2592000 \text{ J} = 2{,}592 \text{ kJ}$$

Now that we have looked at the different forms of energy relevant to an automobile, let us consider *power*, the rate at which energy is used.

The power to accelerate the car from 0-60 mph in 9 seconds (the maximum acceleration for a Corolla) can be calculated as:

$$\text{Power} = \Delta KE/\Delta t = 469.2 \text{ kJ} / 9 \text{ s} = 52.1 \text{ kW} = 69.9 \text{ hp}$$

This is about half of the rated power of the engine.

If a car consumes 30 mpg (12.8 km/L) while driving 60 mph (96.5 kph) then the fuel consumption rate is 2 gal/hr = 7.57 L/hr = 0.00210 L/s.
The density of gasoline is 750 kg/m^3 (SG = 0.75, most liquids are less dense than water), so the mass flow rate of fuel into the engine is (0.00210 L/s)*(0.75 kg/L) = 0.00158 kg/s.
The fuel has chemical energy of 44,000 kJ/kg, so the rate of energy release from burning fuel is:
$$(0.00158 \text{ kg/s})*(44{,}000 \text{ kJ/kg}) = 69.4 \text{ kJ/s} = 69.4 \text{ kW} = 93.0 \text{ hp}$$

For the same Corolla, the power to overcome aerodynamic drag at highway speeds can be calculated from the equation for the aerodynamic resistance force, F_D:

$$F_D = \tfrac{1}{2} C_D A \, \rho \, V^2$$

To use this equation we need to know the drag coefficient of the car (C_D) and its cross sectional area, A (you will learn more about this in your fluid mechanics course). The cross sectional area is 50 by 70 inches, approximately. The equivalent drag area of the car (product of drag coefficient and area) is:
$$C_D = 0.29. \quad C_D A = 6.52 \text{ ft}^2 = 0.606 \text{ m}^2$$

And now we have all the values we need to calculation the drag force, F_D:

$$F_D = \tfrac{1}{2} (C_D A) \, \rho \, V^2 = 0.5(0.606)(1.2)(26.8)^2 = 261 \text{ N} = 58.7 \text{ lbf}$$

The power to overcome drag is the resisting force times the speed at which the object is moved against the force:

$$\text{Power} = F*V = (261 \text{ N})(26.8 \text{ m/s}) = 6999 \text{ W} = 7.0 \text{ kW} = 9.4 \text{ hp}$$

Note 1 N = 1 kg m/s^2, so force times velocity has units of kg m^2/s^3 = kg m^2/s^2 / s = J/s = W

Energy Use in the US

Now let's talk about large-scale energy use. Figure 1 shows the sources and uses of energy in the United States.

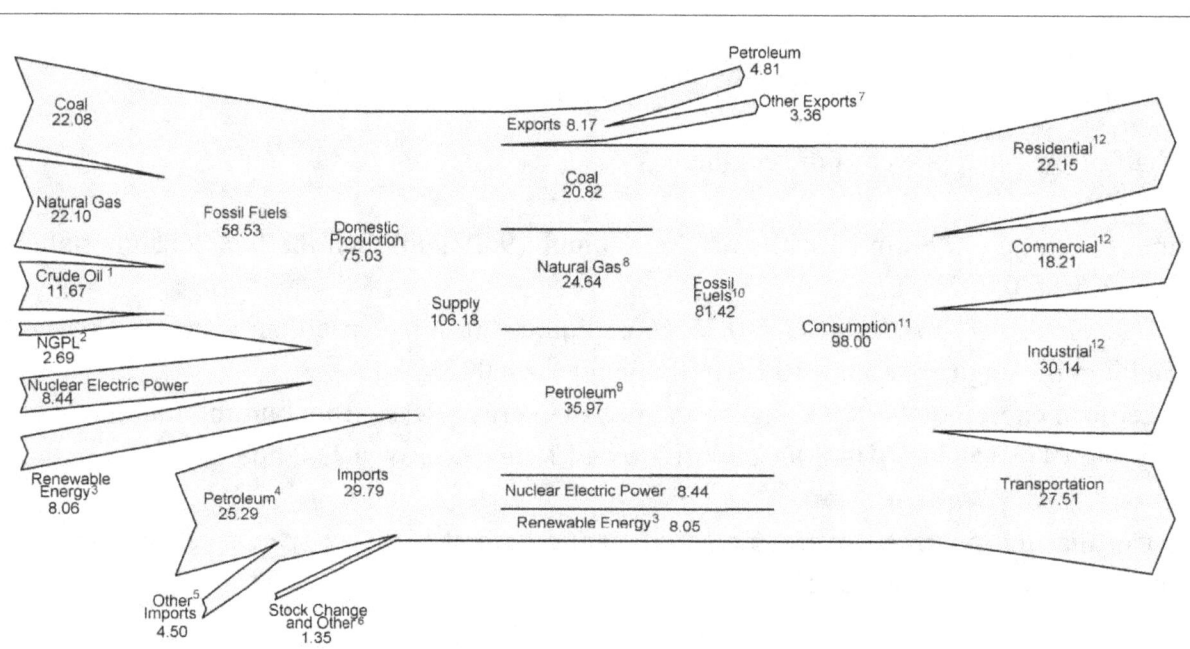

Figure 1: Energy flow diagram for United States in 2010. All numbers in Quadrillion BTUs. From U.S. Energy Information Administration.
https://www.eia.gov/totalenergy/data/monthly/pdf/flow/total_energy.pdf

The "Renewable Energy" arrow in Figure 1 is mostly comprised of hydroelectric power. One of the few examples of **gravitational potential energy** being converted to other forms of energy is the generation of electricity from elevated lakes of water. Hydroelectric power plants make use of the naturally occurring potential energy of water found in rivers when there is a significant elevation change. The change in pressure due to height is given as:

$$\Delta P = \rho g \Delta h$$

and the power generated

$$\dot{W} = \eta \dot{V} \Delta P$$

where η is the efficiency of the turbine, and V is the volumetric flow rate (m³/s) of water. Note: *pumps* and *turbines* will be discussed in more detail later in the section on simple devices.

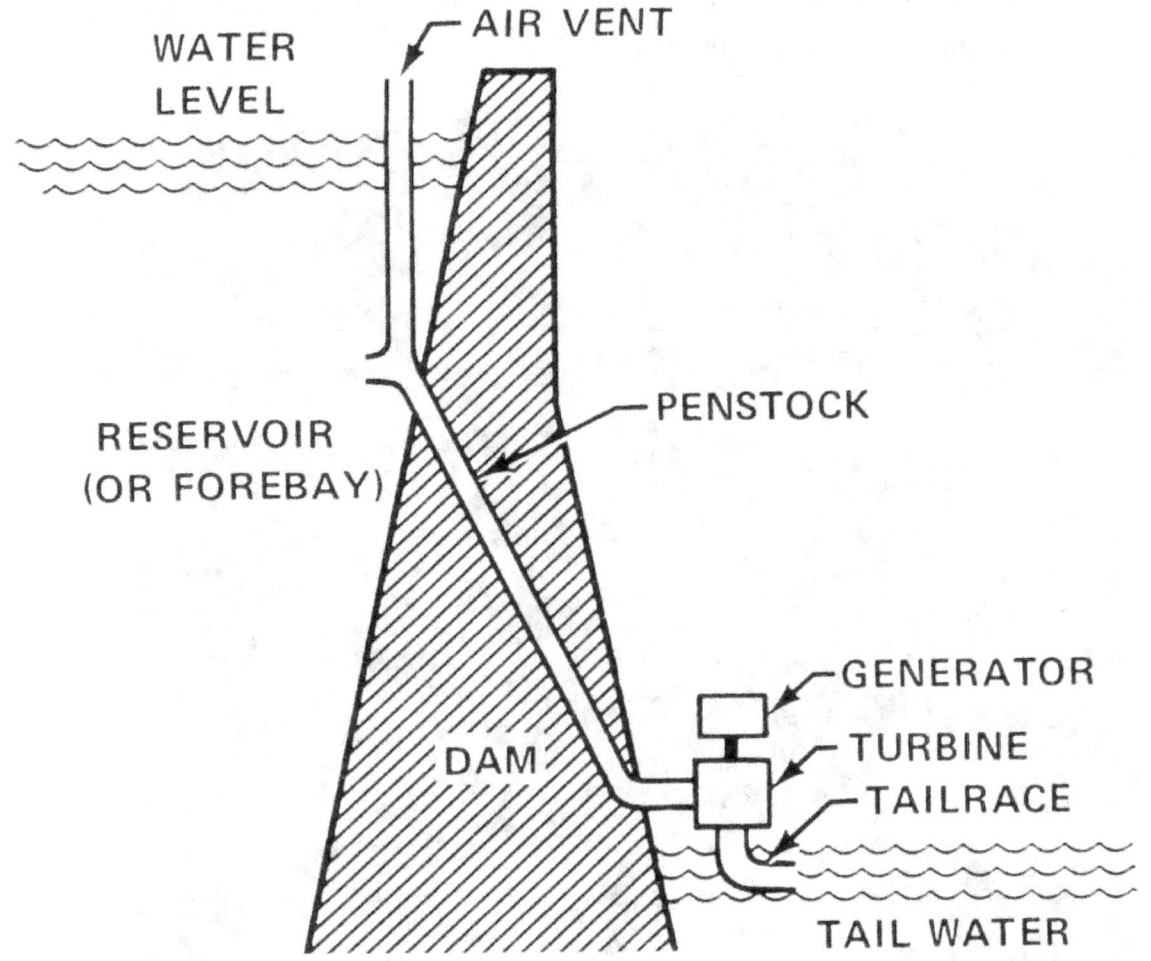

Figure 2: Schematic diagram of hydroelectric power plant. Source [Deskbook]

The three primary types of turbine design used in hydroelectric plants are Pelton, Francis, and Kaplan. Hydroelectric plants can have efficiencies of 80%. The first electric hydropower plant in the U.S. was built in over the Fox River in Appleton, Wisconsin in 1882. [Deskbook] In the United States, there is 80,000 MW of hydroelectric capacity, representing about 10% of U.S. electrical generation, and more than half the total energy that comes from renewable sources. Additionally there is 18,000 MW of pumped storage capacity.

Note: 1 MW = 1 Megawatt = 1000 kW = 1,000,000 watts

The Hoover Dam in Nevada produces about 4 billion kilowatt-hours of hydroelectric power each year, equivalent to an average power of 480 MW year round. The plant has a nameplate capacity of about

2,080 MW. The turbines operate with maximum head of 590 feet (180 m) water elevation, a minimum of 420 feet (128 m), and the average is usually between 510 to 530 feet (155 m to 160 m). The 17 main turbines are Francis-type vertical hydraulic turbines, and it also has two smaller Pelton turbines for plant operations. The maximum flowrate of water through the plant is 49,000 ft³/s = 1390 m³/s, while the average flow rate is about a third of that, at 15,000 ft³/s = 425 m³/s.

Figure 3: The blades and rotor from an old Pelton water wheel. Author's picture.

Example 1:
How much power does the Hoover Dam generate at the average head and average flow rate, assuming an efficiency of 80%?

SOLUTION: The power can be calculated using the following equation:

$$\dot{W} = \eta \dot{V} \Delta P = \eta \dot{V}(\rho g \Delta h)$$

$$\dot{W} = (0.80)\left(425 \ \frac{m^3}{s}\right)\left(1000 \ \frac{kg}{m^3}\right)\left(9.8 \ \frac{m}{s^2}\right)(155 \ m) = 516 \ MW$$

Nuclear - Naturally occurring Uranium is a mixture of about 0.7% U^{235} and 99.3% U^{238}. Uranium-235 is the only naturally occurring fissile isotope, and in sufficient concentration, the radioactive U^{235} isotopes can maintain a sustained nuclear fission reaction, which generates the heat in nuclear power reactors. That heat is the input for a conventional Rankine cycle power plant. For a nuclear fission reactor, the uranium must be enriched so that it contains about 3% U^{235}, while weapons-grade uranium is composed of at least 90% U235. Enrichment is typically done by centrifuges, in which the uranium is first converted to gaseous uranium hexafluoride (UF_6), and the heavier isotopes can be separated from the lighter ones. While the chemical energy contained in petroleum derived fuels such as gasoline is around 40,000 kJ/kg, or 4×10^7 J/kg, the energy that can be released from highly enriched uranium is a million times larger, at 8×10^{13} J/kg. In the U.S., 20% of electricity comes from nuclear power, totaling about 100,000 MW from 104 reactors.

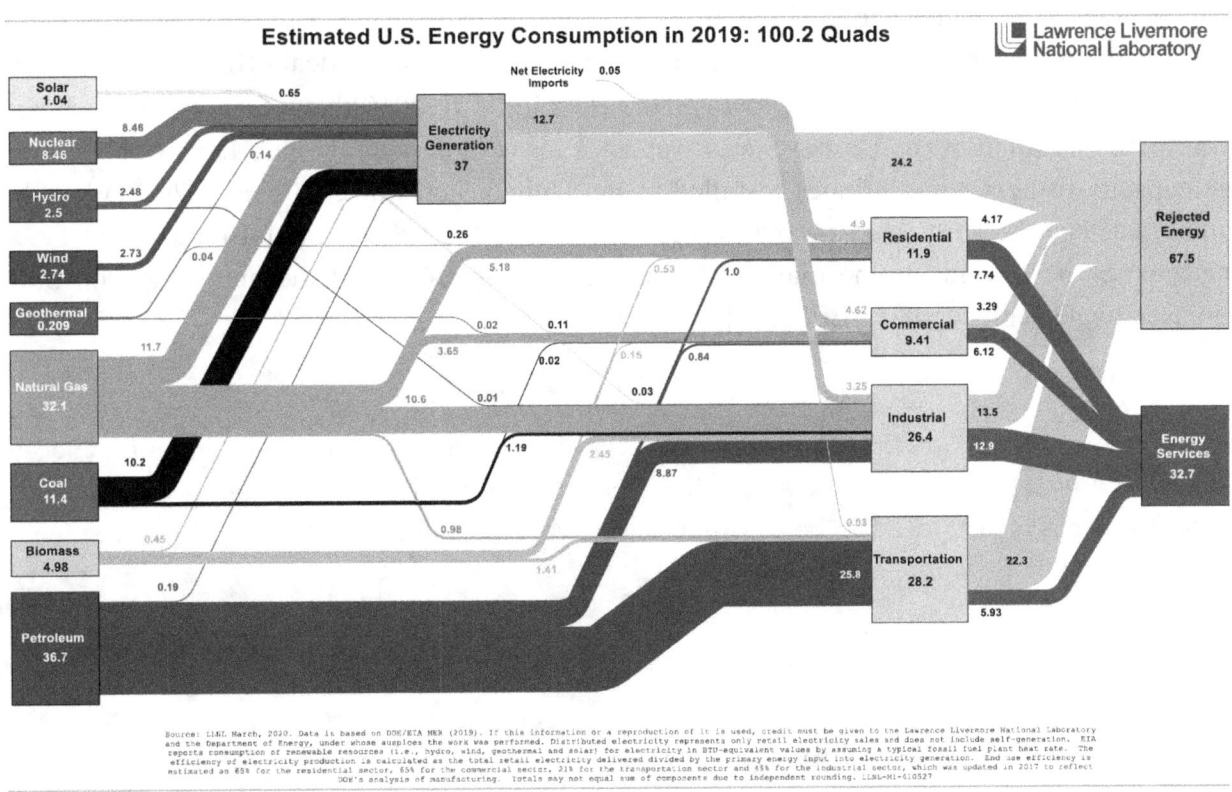

Figure 4: Energy flow diagram. From Lawrence Livermore National Laboratory.
https://flowcharts.llnl.gov/commodities/energy

Wind Energy

The energy that a wind turbine extracts from the wind is the form of the kinetic energy (KE) of motion of the air. The kinetic energy per unit mass is given by:

$$\frac{KE}{m} = \frac{V^2}{2}$$

and the mass flow rate is:

$$\dot{m} = \rho A V$$

So that the mechanical power that can be generated by a wind turbine is the product of the flow rate and the kinetic energy per unit mass of the air, multiplied by the efficiency of the turbine in converting in converting kinetic energy to electrical energy:

$$\dot{W} = \eta \frac{1}{2} \rho \left(\frac{\pi}{4} D^2\right) V^3$$

where η is the efficiency of the turbine, ρ is the air density, which is primarily a function of altitude, D is the diameter of the rotor, and V is the wind velocity. This maximum theoretical efficiency of a wind turbine is η_{max} = 0.593, or 59.3%, which is known as the Betz limit. In 2010 the average rotor diameter of the wind turbines installed in the United States was 84.3 m, with an average hub height of 79.8 m. The average capacity of new wind turbines installed in the United States in 2010 was 1.79 MW each, and that number has been steadily growing over time, with a trend for larger wind turbines. From 1998 to 2010, the average new wind turbine diameter size increased by 76% and the hub height on new turbines increased by 43% on average. [EERE11]

Example 2:
A typical residential home uses on the order of 1 kW electrical power on average, not counting energy used for heating. If a home is built in a location where the average wind speed is 4 m/s (about 9 mph), how big would the wind turbine need to be to provide the average electrical power needed? Assume a turbine can be purchased which has an efficiency of 30%.

SOLUTION: The power generated by a wind turbine is given by:

$$\dot{W} = \eta \frac{1}{2} \rho \left(\frac{\pi}{4} D^2\right) V^3$$

This can be rearranged to solve for the diameter of the wind turbine:

$$D = \sqrt{\frac{8\dot{W}}{\pi\eta\rho V^3}} = \sqrt{\frac{8(1000\ W)}{\pi(0.3)(1.2\ kg/m^3)(4\ m/s)^3}} = 10.5\ m$$

This is equivalent to about 35 ft in diameter, which will require a tower at least 50 ft high. Note this reflects only the *average* wind power needed. Since the winds fluctuate, a way of storing excess energy or a backup power source would be needed to make this home truly grid independent.

Figure 5: Old 2-bladed wind turbine design. From NASA Glenn Image Archive.

Solar power refers to use of the radiation arriving at the earth from the sun to create electricity. The solar constant (the amount of energy reaching the upper atmosphere from the sun) is 1.353 kW/m². On a clear sunny day the amount of energy reaching the earth's surface is about 1.0 kW/m². Solar energy reaching the earth's surface is 46% visible light, 46% infrared, and 8% ultraviolet. [Deskbook] Information about solar radiation fluxes at different sites on the ground in the United States can be found at the National Solar Radiation Data Base at: http://rredc.nrel.gov/solar/old_data/nsrdb/ with data available in monthly and hourly averages. Additional data is also available at http://www.nrel.gov/rredc/solar_data.html

Figure 6: Solar-powered unmanned aircraft. From NASA Armstrong Image Gallery.

<u>Energy Storage</u>

Methods of **energy storage** include pumped hydroelectric, compressed air, hydraulics, flywheels, chemical (if reactions are reversible), electromagnetic, thermal, batteries, fuel cells, and capacitors. A popular way to compare energy storage options is to make a *Ragone Chart*, which plots *specific power*, usually in units of W/kg, vs. *specific energy*, usually in units of W-hr/kg. For large-scale energy storage from the electrical grid, the only two technologies in current implementation are compressed air energy storage (CAES) and pumped hydroelectric energy storage (PHES). Energy storage can improve the usability of wind energy, which suffers from the problem that the daily peaks in wind power do not coincide with the peaks in demand for electricity. In some locations, about two-thirds of the wind power produced is outside of the times of peak electrical demand [Baxter06]. Flywheels have seen applications for local short-term emergency power, and hydrogen and batteries are of interest for vehicles. Currently in the United States there is 22 GW of capacity at 150 facilities for pumped hydroelectric storage of electrical energy, 110 MW at one facility for compressed air energy storage, more than 70 MW of capacity of battery storage at multiple sites ranging up to 20 MW, and 100 MW in superconducting magnetic energy storage, with individual sites up to 10 MW. [EPRI03].

As of 2018, there was 25.2 GW of energy storage capacity in the United States, 94% of which is pumped hydroelectric. Of the remaining, 733 MW is electrical batteries, 669 MW is thermal storage, 114 MW is compressed air, and 58 MW is flywheel. (Source: https://www.epa.gov/energy/electricity-storage)

Compressed Air
Compressed Air Energy Storage (CAES) uses off-peak electric power to compress air, which is stored in a reservoir underground. There are currently two operating CAES facilities. The first was built in Huntorf, Germany in 1978. It is rated at 290 MW, and can generate that much power for 4 hours. It uses an underground salt cavern of 280,000 m^3 capacity, with pressures up to 6.9 MPa (1000 psi). It has an average 90% availability. The compressors have an efficiency of 83%. [Baxter06] At full power, an air flow rate of 417 kg/s is used. [EPRI03] The McIntosh, Alabama CAES facility was built in 1991. It also uses an underground salt mine, of 540,000 m^3 capacity, with pressures up to 7.6 MPa (1100 psi). At maximum charging, it can provide 100 MW for 26 hours. [Baxter06] At full power, an air flow rate of 155 kg/s is used. [EPRI03] Efficiency of compressed air storage and release is also estimated to be around 70%. [Deskbook] CAES facilities are capable of black start and have fast startup times, as little as 5 minutes. [EPRI03]

Hydroelectric Pumped Storage
Pumped Hydro-Electric Storage (PHES) is also called Pumped Storage Hydropower (PSH). A typical PHES facility consists of two water reservoirs at different elevations, with a reversible pump/turbine between them. During off-peak hours excess electricity from the grid is used to pump the water from the lower reservoir to the upper reservoir, and then during peak hours water is allowed to flow back down to the lower reservoir, generating power in the turbine. For modern facilities the round trip efficiencies are around 75%. [Baxter06] At least 300 m of elevation difference is generally preferred in PHES facilities. PHES facilities can go from complete shutdown to full power in a manner of minutes. The availability of PHES facilities is over 98%. The losses in PHES include inefficiencies in the pump/turbine unit, and the friction losses in pumping water through the pipes between the two reservoirs. Typical size of PHES facilities is from 300 MW to 1800 MW. There are more than 240 PHES facilities worldwide. [Baxter06] Pumped Storage facilities store off-peak power (such as generated by nuclear plants, that usually run at constant capacity), and then provide reliable generating during peak hours. They also provide system load regulation, rapid response to sudden load fluctuation, and system voltage regulation, which contributes to overall system reliability. [Hydro96] They also have black start capability, meaning they are able to restart even with a system-wide electrical transmission failure. Because they do not require external power to startup, they can be used to provide power in even of a blackout. Rankine cycle power plants powered by coal or nuclear fuel do not have black start capability, because they require significant amounts of electrical power to run the water pumps and other accessories. Modern PHES can go from shutdown to full load generating in about 2 minutes, compared to several hours required for a coal plant to come to full power after shutdown. [Hydro96]

As of 2010, there is 127,000 MW of pumped storage capacity worldwide. There is about 21,000 MW of pumped storage capacity in the United States, about 2.5% of total generating capacity. PHES provides enhanced stability and reliability of the electrical power system, and provides off peak storage and regulation that can be used to accommodate increasing use of renewable energy sources that have high short-term variability (solar and wind). [PSH10] The leading nations in terms of pumped storage capacity include Japan, the United States, Italy, Germany, France, Spain, and China.

Examples of PHES
- The Grand Maison Pumped Storage Plant was built in the French Alps in the 1980s. It has an installed capacity of 1800 MW and operates with an average hydraulic head of 955 m. Full flow is 217 m^3/s. with usable storage of 14 Mm^3. [Hydro96]
- The Dinorwig Pumped Power Station in Great Britain has an installed capacity of 1800 MW, energy storage of 8400 MW-hr, overall cycle efficiency of 78%. [Hydro96] It can produce 1800 MW for up to 5 hours. 6 million cubic meters of water, with a head of 600 m. When in spinning reserve mode, the plant can reach full power in 16 seconds. [Baxter06]
- The Taum Sauk power station, rated at 450 MW, is located in Missouri. The plant resumed operations in 2010 after 2005 upper reservoir wall breach. In 2005 an overtopping accident caused the upper reservoir wall to fail, releasing the entire contents of the upper reservoir of 1.4 billion gallons of water (5,363,000 m^3) in 25 minutes, with a peak flow of 270,000 ft^3/s (7,650 m^3/s). The change in elevation between the upper and lower reservoirs varies between 776 ft (236 m) and 860 ft (262 m). It has a maximum operating flow rate of 3000 ft^3/s (85 m^3/s). [FERC06] During a typical 24-hour period of operation at Taum Sauk, pumping to the upper reservoir begins around 10:00 pm as excess power from the grid becomes available. Pumping continues on through the night until about 6:00 am as either the upper reservoir limit level is reached or excess grid power is no longer available. From 6:00 am to noon the base load plants are typically able to supply the power demands of the electrical grid, so Taum Sauk is usually idle at this time. Generation of power at Taum Sauk begins around noon and continues for about 5 hours. Generation stops by 6:00 pm as the demand for power ebbs. [FERC06] The passageways for pumping water between the two reservoirs consists of a 27.2 ft (8.3 m) diameter vertical shaft, with a length of 451 ft (137 m), a 25 ft (7.6 m) diameter unlined horseshoe tunnel sloping at 5.7% grade over a length of 4765 ft (1453 m), and a horizontal steel lined tunnel of diameter 18.5 ft (5.6 m) and length 1807 ft (551 m). [FERC06]

Thermal energy storage (TES) – Energy can be stored thermally in a fluid that can be moved when needed. Some TES systems work with a building's cooling system, and produces chilled water (or ice or a chilled water/ethylene-glycol system) using electricity from the grid during off-peak hours and storing the cooled product in an insulated tank. During the day when heating loads and electrical loads are high, the cooled fluid is used to provide at least a portion of the building's cooling load. [Baxter06]

Enthalpy

To calculate how much energy we can store thermally in a working fluid, we need to be able to calculate how much internal energy (u) a fluid can store as a function of temperature (T). The specific heat capacity is a measure of a substance's ability to store energy thermally. The *constant pressure specific heat capacity* (c_P) and *constant volume specific heat capacity* (c_V) are defined as:

$$c_P = \frac{dh}{dT}$$

$$c_V = \frac{du}{dT}$$

where h is the **enthalpy**. Both c_P and c_V have units of kJ/kg-K. The definition of enthalpy is:

$$\boxed{h = u + Pv}$$

This is simply a definition. There is nothing intrinsically special about enthalpy, but the term "Pv" occurs so often in engineering problems engineers have found it convenient over the years to combine it with internal energy and named the sum the enthalpy. For solids and liquids specific heat is nearly a constant, so that the derivative can be replaced by a difference:

$$c_P = \frac{dh}{dT} \approx \frac{\Delta h}{\Delta T}$$

Further, for solids and liquids, the specific volume, v, is typically very low (for water at STP $v = 0.001$ m³/kg) and does not change much. To good accuracy, we can often model liquids as an *incompressible* fluid, where the density (and specific volume) do not change. With this assumption we can write:

$$\frac{d}{dT}(Pv) = 0$$

and

$$c_P \approx c_V = c$$

and only a single value of heat capacity, c, is used. For liquid water at STP, c = 4180 J/kg °C. (Note: 1 kilocalorie = 4.18 kJ).

Water is limited to changes of less than 100 °C at atmospheric pressure to remain in the liquid state. So the maximum possible energy density when using water for thermal energy storage is about:

$$h = c_P \Delta T = (100 \text{ °C})(4.18 \text{ kJ/kg-K}) = 418 \text{ kJ/kg} = 116 \text{ W-hr/kg}.$$

Du Pont's heat transfer salt (HTS) compound of 49% $NaNO_3$, 44% KNO_3, and 7% $NaNO_2$ melts at 142 °C and can be used up to 540 °C, with a heat capacity of 1.55 kJ/kg-K and density of 1760 kg/m^3. So the maximum possible specific energy density for this substance would be

$$h = c_P \Delta T = (540 \text{ °C} - 142 \text{ °C})(1.55 \text{ kJ/kg-K}) = 617 \text{ kJ/kg} = 171 \text{ W-hr/kg}.$$

For gases the heat capacities do vary significantly with temperature, and the specific volume, v, changes with both pressure and temperature. Many thermodynamic properties vary much more strongly with temperature than pressure. The number of properties needed to define a thermodynamic state will be discussed in the section on the state principle and Gibbs Phase rule. The simplest model for a gas is the *ideal gas law*. For an ideal gas recall that:

$$Pv = RT$$

The ideal gas law will be covered in more detail later in this book (one challenge of thermodynamics is that so much is inter-related, it is difficult to cover each topic in a linear sequential order.)

So we can use the ideal gas law to substitute $pv = RT$ into the definition of enthalpy ($h = u + Pv$), so that $h = u + RT$. Since $c_P = dh/dT$ and $c_V = du/dT$, for an ideal gas:

$$c_P = c_V + R$$

For gases c_P and c_V both vary with temperature (the explanation for this requires some quantum mechanics), as shown in Figure 7.

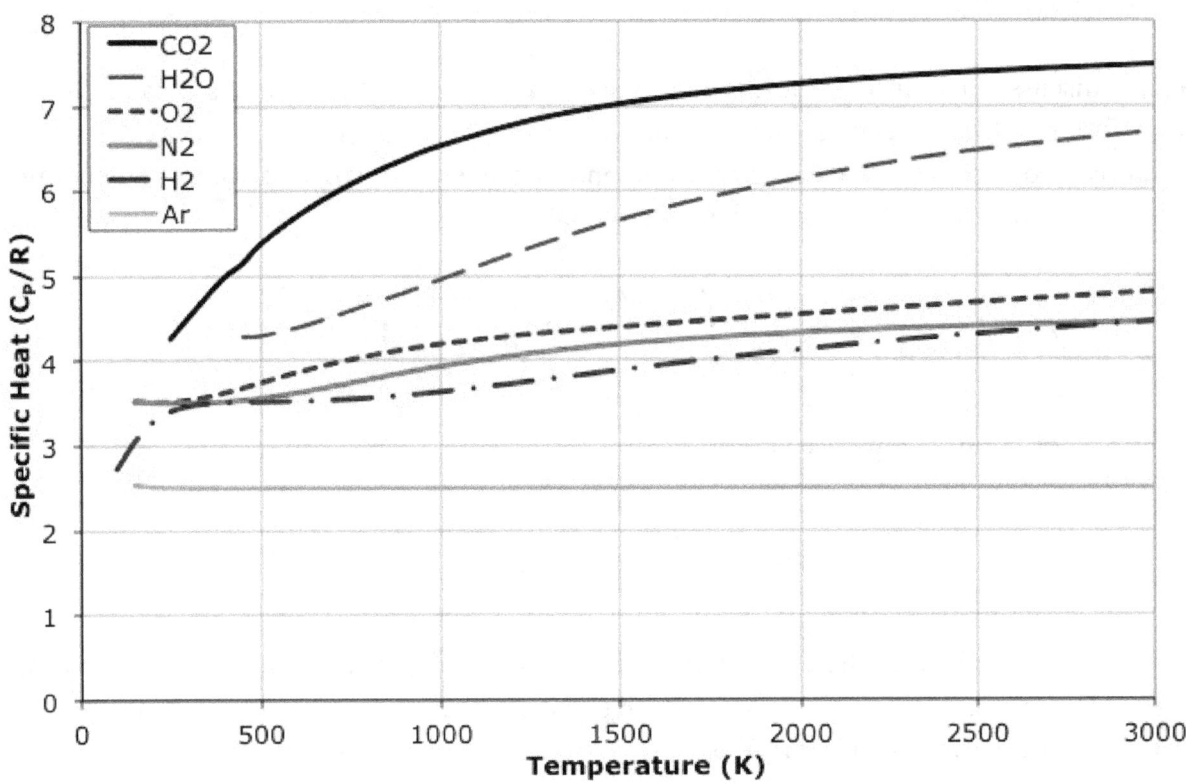

Figure 7: Constant pressure heat capacities for selected triatomic, diatomic, and monatomic gases.

Table 1: Energy storage densities for different types of technologies.

Technology	Energy density (kJ/kg)	Energy density (kJ/L)
Pumped hydro (Δz = 100 m height)	0.98	0.98
Thermal storage (water ΔT = 100 °C)	418	418
Compressed Air (300 bar)	500	170
Battery (lithium ion)	700	1400

First Law of Thermodynamics

Now that we have discussed the different forms energy can take and the different ways it can be stored, it would be a good time to write a general formula for *conservation of energy*, also know as the **1st Law of thermodynamics**. For a system containing mass, m, and total energy E, we allow mass and energy to flow into and out of the system, with the following energy balance:

$$\frac{dE}{dt} = \sum \pm \dot{Q} + \sum \pm \dot{W} + \sum_{in} \dot{m}\left(h + \frac{V^2}{2} + gz\right) - \sum_{out} \dot{m}\left(h + \frac{V^2}{2} + gz\right)$$

where *E* is the energy of the *system* we are analyzing:

$$E = me = m\left(u + \frac{V^2}{2} + gz\right)$$

The variable symbols are defined in the table below. The ± symbol in front of the work (W) and heat transfer (Q) terms denotes that the energy flow is positive if it adds energy to the system and negative if it extracts energy from the system. In total, there are 9 terms in the conservation of energy equation above, but as you will see in the subsequent problems in this book, usually we can eliminate most of the terms that are either exactly 0 or close enough to 0 to be ignored for a particular problem.

Variable	Name	Importance	Units
E	Energy of the system	Total energy	J
e	Specific energy	Energy per unit mass	J/kg
t	Time	Transient changes	s
$\dot{Q}$	Heat Transfer	Movement of thermal energy	J/s = W
$\dot{W}$	Work	Mechanical Energy	J/s = W
m	Mass		kg
$\dot{m}$	Mass flow rate		kg/s
u	Internal Energy	Internal thermal energy	J/kg
h	Enthalpy	Internal thermal energy + pressure energy	J/kg
V	Velocity	Kinetic energy	m/s
g	Acceleration due to gravity	Gravitational potential energy	m/s²
z	Height above reference line	Gravitational potential energy	m

Waste Heat

When we burn a fuel in a combustion process to do work (such as move the pistons in a car engine or spin the turbines in a power plant) we can never extract 100% of the energy do work. There is always some heat that is wasted. Why is the exhaust coming out the tailpipe of your car hot? Because not all of the energy released from the fuel burn was able to be captured in the piston motion. Your car also looses heat through the radiator. The *rule of thirds* is a decent approximation for the heat balance in a gasoline engine. Roughly 1/3 of the fuel energy going into the engine goes out the tailpipe, 1/3 goes out the radiator, and 1/3 does something useful – move the car (and power the accessories, like lights). Thermal power stations (particularly those that are coal or nuclear powered) typically have big cooling towers with large white clouds coming out the top? Why are they needed? The same reason – there is waste heat that must be expelled to the environment in some manner.

Why is there waste heat? One way to answer that question is the *2nd law of thermodynamics*. Unlike the first law of thermodynamics (conservation of energy) which has one clear statement (energy cannot be created or destroyed, but only transferred from one form to another), the 2nd law of thermodynamics is expressed in a few different ways.

- Heat is always naturally transferred from a hot object to a cold one (unless external work is being done, as in a refrigeration system)
- Heat cannot be converted completely to mechanical energy (there must always be some waste heat in any process)
- The total entropy of a thermal-mechanical system and its surrounding environment can not decrease, though it can remain constant in a reversible process.

Entropy

The second law also introduces us to a new thermodynamic property, **entropy**. Entropy has a variety of physical interpretations, including the statistical disorder of the system on the molecular level, but for our purposes, let us consider entropy to be just another property of the system, like enthalpy or temperature. In its equation form, the second law of thermodynamics tells us that in a thermodynamically **reversible** process, the change in entropy (ΔS) is equal to the amount of heat transferred (ΔQ) divided by the absolute temperature (T):

$$\Delta S_{rev} = \frac{\Delta Q}{T}$$

As with ideal gas law, T must be in absolute units (Kelvin or Rankine). Like with many other thermodynamic properties, the total entropy, S, is related to intensive or specific entropy, s, through the mass (m):

$$S = m\,s$$

thus entropy, **s**, has units of kJ/kg-K.

For an **adiabatic** process, $\Delta Q = 0$. Thus for a process that is both *adiabatic* and *reversible*, $\Delta S = 0$. Such a reversible adiabatic process is thus called an **isentropic** process, as the entropy will be constant.

What is a reversible process? It is one that can be returned to its initial state without any net loss of energy. In reality all processes have some degree of irreversibility, but it is often useful to *model* certain processes as reversible. What are the reasons why real processes are irreversible? These include friction, turbulence, and molecular diffusion, which cause either the transfer of work to heat or the mixing of substances in such a way that they can not naturally be reversed, but would require external work to be done to put the system back in its original state. Note it is often easy to convert other forms of energy to heat. The reverse is not always easy. Why do you think that is? Another way to think of entropy is that it is the *unavailability* of heat transferred from the system to do mechanical work.

What is mechanical work? From basic physics, recall that work is equal to the product of force (F) times the distance over which the force acts (Δx).

$$W = F\,\Delta x$$

Now this equation is only valid if the force, F, is a constant during the process. More generally, if the force changes over time, then we need to use integral calculus:

$$W = \int F\,dx$$

Force is also equal to pressure times the area over which the force acts:

$$F = P\,A$$

In a thermo-mechanical process, the differential change in volume, dV, is equal to the cross-sectional area (A) times the differential change in distance perpendicular to the area, so that:

$$dV = A\,dx$$

Substituting this into the integral equation for work, we can express mechanical work in terms of thermodynamic properties:

$$W = \int P \, dV$$

For pressure in Pascals and volume in m³, this equation will give work in Joules. On a per-mass basis, you would use the specific volume, v:

$$\frac{W}{m} = \int P \, dv$$

Heat Transfer

Heat flows from hot to cold just as fluids go from high to low pressure in a pipe and water flows down hill due to gravity. In fact, heat is always transferred when there is a temperature difference between two objects or fluids in close proximity. There are three mechanisms, or modes, by which heat can be transferred, summarized by the poem:

Some like it hot,
some like it cold,
some like heat transfer,
in all three modes

The three modes are *Conduction, Convection,* and *Radiation*.

Conduction is the transfer of heat through direct contact in a material. It is calculated by:

$$Q = kA \frac{dT}{dx}$$

where k is the *thermal conductivity* of the material, having units of [W/m-K], A is the cross-sectional area through which heat is being transferred, and dT/dx is the temperature difference over a given distance. Materials of high thermal conductivity transfer heat readily and are referred to as good conductors (most metals have high values of k), while materials of low thermal conductivity make good *insulators*.

Convection is the transfer of heat from a solid object to a fluid (liquid or gas) passing over that object. The equation for convective heat transfer is:

$$Q = h * A * (T_{gas} - T_{cool})$$

Where:

Q = overall heat transfer (W)
A = reference cylinder area (m^2)
T_{gas} = effective gas temperature, computed in your MATLAB code
T_{cool} = coolant temperature, typically 80 C
h = heat transfer coefficient (W/m^2 K)

Typically there is no theoretical equation for h, but it must be found using empirical correlations (meaning curve fits to experimental data).

Radiation is the transfer of heat through electromagnetic radiation. Unlike conduction and convection, which rely on molecule to molecule contact to transfer energy, in radiation the two bodies exchanging energy need not be touching. The equation for radiation heat transfer from an object to the environment is:

$$Q = \varepsilon \sigma A (T_s^4 - T_e^4)$$

Where σ is the *Stephan-Boltzmann constant*, which has a value of 5.6703 10^{-8} (W/m²K⁴), and ε is the *emissivity* of the surface [non-dimensional] which depends on the material and the finish of the surface, A is the surface area[m²], and T_s and T_e are the temperature of the surface emitting the radiation and the temperature of the surrounding environment receiving the radiation, respectively. As with the ideal gas law, the temperature must be in absolute units (Kelvin).

Figure 8: Mach-3 capable SR-71B blackbird. From NASA Armstrong Image Gallery.

Example 3:
How hot can the surface temperature of the SR-71 get when it is cruising at Mach 3.2 at an altitude of 25,000 m (82,000 ft)? How much power does it radiate out when cruising?

At an altitude of 25,000 m, the U.S. Standard Atmosphere Tables (can be found online, such as at: https://www.engineeringtoolbox.com/standard-atmosphere-d_604.html) gives relevant properties of air as:
- Pressure of 0.0255 bar
- Temperature of -51.6 °C = 221.5 K
- Density of 0.04 kg/m^3
- Speed of sound of 298 m/s

SOLUTION: A general form of the First Law of thermodynamics is:

$$\frac{dE}{dt} = \sum \pm \dot{Q} + \sum \pm \dot{W} + \sum_{in} \dot{m}\left(h + \frac{V^2}{2} + gz\right) - \sum_{out} \dot{m}\left(h + \frac{V^2}{2} + gz\right)$$

We can simplify the First Law with the following observations and assumptions:
- For this problem, we are considering a steady-state situation (flight at constant speed and altitude), so dE/dt = 0.
- Since the altitude does not change $z_{in} = z_{out}$.
- No mechanical work is being done outside the plane, so W = 0

Recall that *enthalpy*, h = u + Pv, or *internal energy* (u) + pressure energy.

In steady state, no work or heat transfer, no change in potential energy, the 1st Law simplifies to:

$$\left(h + \frac{V^2}{2}\right)_{in} = \left(h + \frac{V^2}{2}\right)_{out}$$

At a stagnation point on the front of the plane, the velocity slows to V = 0 (this will be the hottest point on the surface). As an approximation, we can assume a "perfect gas", defined as an ideal gas with constant specific heats. Then we can write the change in enthalpy as:

$$h_2 - h_1 = c_P(T_2 - T_1)$$

Combining the two equations, we can solve for the unknown T_2:

$$T_2 = T_1 + \frac{V^2}{2c_P}$$

So for cruising condition at Mach 3.2, V = 3.2*298 m/s = 955 m/s (3430 km/hr or 2130 mph)
We can now substitute in values to solve for T_2:

28

$$T_2 = 221.5\ K + \frac{(955\ m/s)^2}{2(1005\ J/kg \cdot K)} = 221.5\ K + 453.7\ K = 675\ K$$

This is equivalent to 402 °C or 756 °F.

This is the temperature at the hottest part of the aircraft, at the nose and leading edges. The *average* temperature of the aircraft surface is cooler, around 260 °C, and obviously the internal temperature of the cockpit must be kept at a reasonable temperature for its human occupants.

To calculate the heat transfer, we need to know the surface area (contact area for heat exchange) of the aircraft. The dimensions of the SR-71 are: length 32.7 m, wingspan 16.9 m, and height 5.6 m (including landing gear). The total effective radiative area is ~ 400 m². The SR-71 is painted black to increase its surface emissivity (ε = 0.9) to increase the radiative heat transfer cooling. Assuming an environment temperature of 280 K, we can calculate the radiation heat transfer as:

$$Q = \varepsilon \sigma A (T_s^4 - T_e^4)$$

$$Q = 0.9 * 5.67\ 10^{-8} * (400\ m^2) * (533^4 - 280^4) = 1520\ kW$$

Note that just as for ideal gas law, radiation calculations must always use absolute temperature (surface temperature 260 °C = 533 K)

Thermodynamic Properties & Phases of Matter

For a pure substance, we need two pieces of information (such as thermodynamic properties) to completely define the thermodynamic state. As a practical matter, temperature, pressure, mass, and volume and the quantities that are usually the easiest to measure.

Table 2: Commonly used thermodynamic properties

Name	Symbol	Units
Internal Energy	u	kJ/kg
Enthalpy	h	kJ/kg
Entropy	s	kJ/kg-K
Specific Volume	v	m³/kg
Density	ρ	kg/m³
Gibbs Free Energy	g	kJ/kg
Helmholtz Free Energy	a	kJ/kg
Pressure	P	Pa
Temperature	T	K

Density is the mass of a fixed amount of a substance divided by the volume it takes up. The Greek letter rho (ρ) is used for density, and it has units of kg/m³ in standard units.

$$\rho = \frac{m}{V}$$

Specific volume is the reciprocal, or inverse, of density, denoted by an italic v or the Greek letter nu (ν).

$$v = \frac{1}{\rho} = \frac{V}{m}$$

If you know one of specific volume or density you can easily calculate the other. In my career as an engineer I have always used density in my calculations and analysis, but specific volume seems to be default variable used in thermodynamics textbooks, and is probably being used in the course you are taking right now.

Specific gravity (SG) is the density of a substance divided by the density of a reference fluid. For liquids the reference fluid is liquid water at standard conditions and for gases the reference fluid is air. The density of water at T = 20 °C and P = 1 atm is ρ = 1000 kg/m³ = 1 g/cm³ = 62.4 lbm/ft³. A pressure of 1 atmosphere is 1 atm = 101,350 Pa = 101.3 kPa = 1.013 bar = 14.7 psi, where in English units a psi is a pound per square inch of pressure, 1 psi = 1 lbf/in² = 144 lbf/ft².

Absolute pressure is pressure relative to a total vacuum, where P = 0. **Gage pressure** is the pressure relative to local atmospheric pressure, where P = P$_{atm}$ at sea level. Absolute pressure is always positive, while gage pressure can be positive or negative. When you measure tire pressure or pressure in bike tires or in a sports ball, that is a gage pressure, while the pressures in thermodynamic data tables are usually absolute pressures.

Table 3: States of Matter

State	Definition
Solid	Fixed volume, holds shape
Liquid	Fixed volume, amorphous
Gas	Expands to fill volume of container
Supercritical fluid	$P > P_c$ and $T > T_c$, volume can vary
Plasma	Gas-like mixture of neutral atoms, electrons, and positively charged ions
Bose-Einstein Condensate	T close to 0 K (absolute zero)

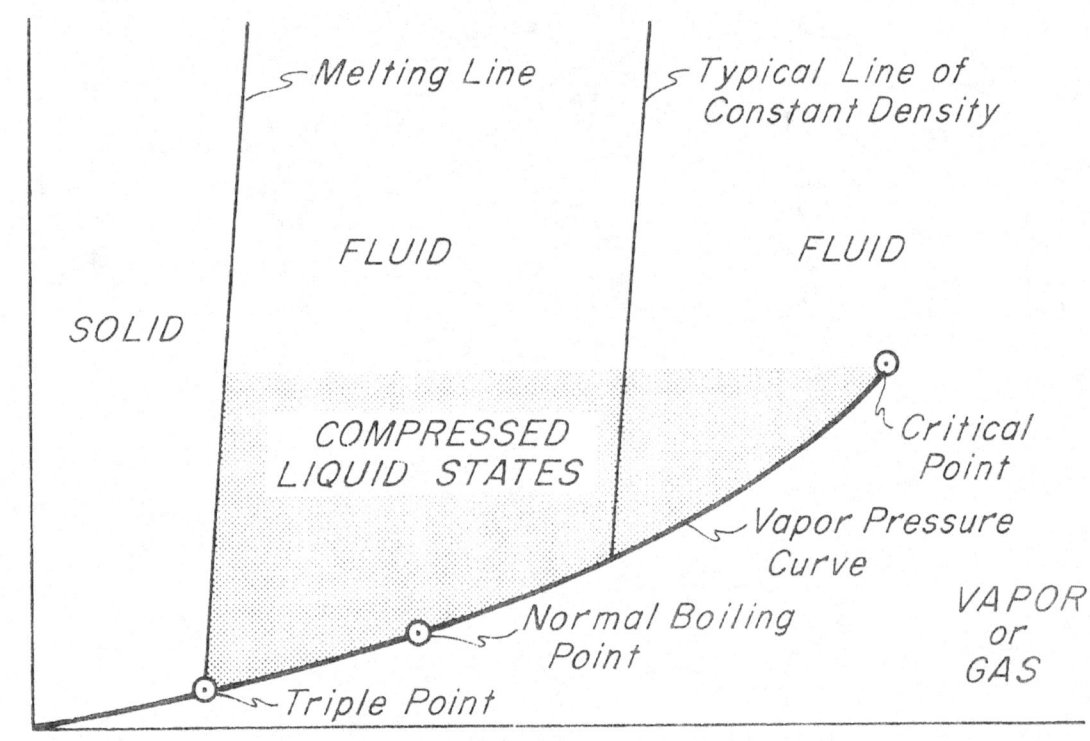

Figure 9: P-T diagram showing states of matter. [NBS74]

If a two-phase mixture of liquid and vapor is held in a constant volume vessel, the denser liquid will settle to the bottom and a meniscus is observed at the interface between the two phases. If the vessel is heated so that the temperature and pressure rise, and the mixture eventually passes into the supercritical region, then the meniscus will disappear and only one continuous phase will be observed. As the supercritical region is approached, the interface boundary between gas and liquid becomes thicker. The critical point can be determined graphically by plotting isotherms on a P-v diagram. At the critical point $dP/dV = 0$, and $d^2P/dV^2 = 0$. The critical point could also be determined by plotting the densities of the saturated liquid and saturated vapor vs. temperature. The point where the slope of the curves is infinite is the critical point, as shown in Figure 12. Applications of supercritical fluids include the use of cryogenic fluids and operation of steam power plants that pass into the supercritical regime. T_b is the *normal boiling point*, the boiling point at a pressure of 1 atm.

Table 4: Critical Points of some common substances. (data from Appendix Table A1)

Substance	Critical Temperature (K)	Critical Pressure (bar)	Critical Density (kg/m^3)	Triple point (K)
Water	647	221	322	273
Ammonia	405	113	235	195
Carbon Dioxide	304	73.8	468	217
Methane	191	46.1	163	91
Oxygen	155	50.5	436	54
Nitrogen	126	33.9	314	63
Hydrogen	33	13.0	31.4	14
Helium	5	2.27	69.9	2.2

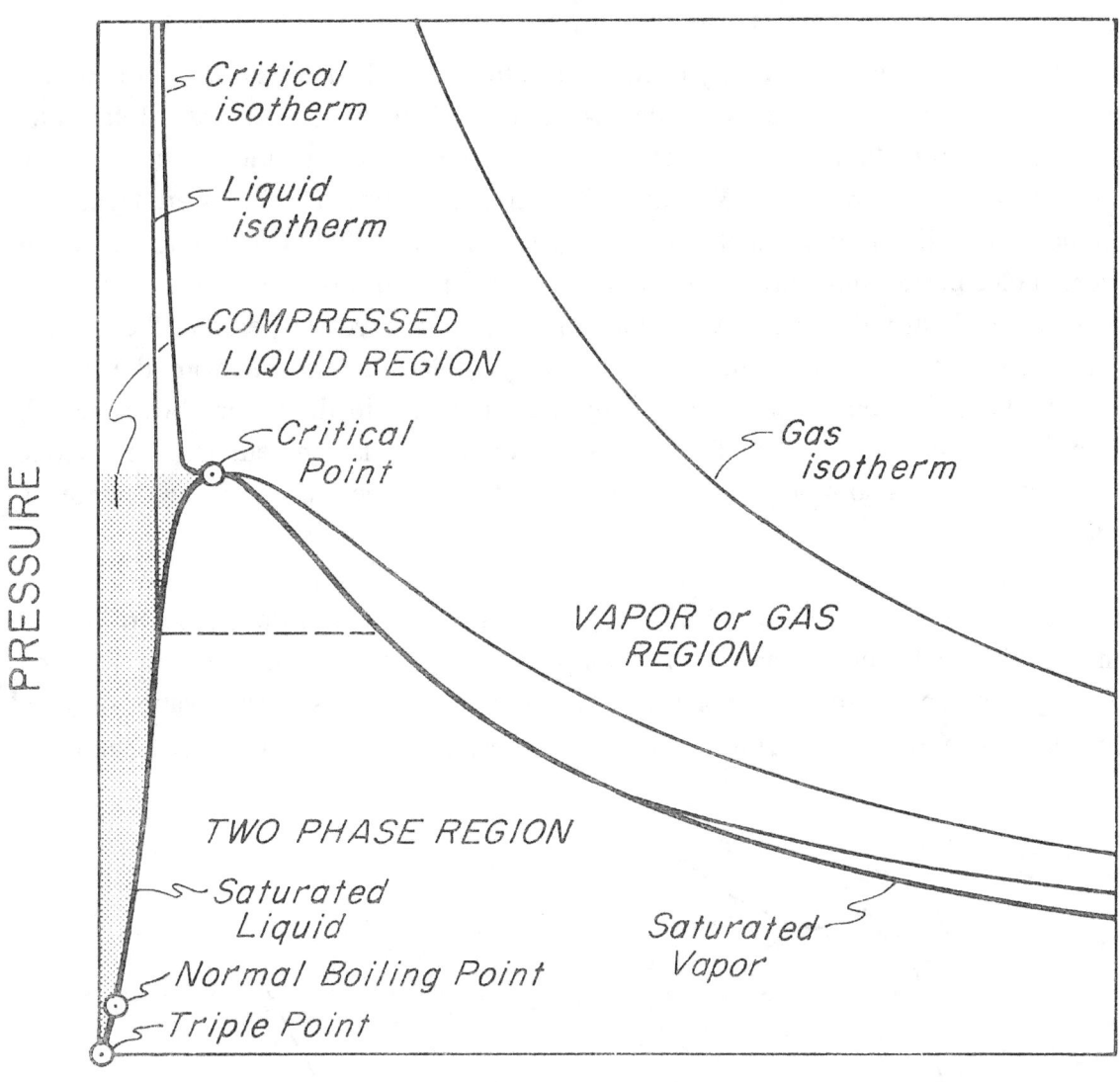

Figure 10: P-v diagram showing states of matter. [NBS74]

In the two-phase region (also called the vapor dome) in Figures 10 and 11, the pressure and temperature curves are flat, completely horizontal. So we need an additional parameter to define where we are on that line when there is a mixture of saturated liquid and saturated vapor. This variable has been defined as the **quality**, x. The quality is the fraction of vapor in the two-phase mixture, on a per mass basis. Mathematically it is defined as:

$$x = \frac{m_{vapor}}{m_{total}} = \frac{m_{vapor}}{m_{liquid} + m_{vapor}}$$

Quality is non dimensional and can have values from 0 to 1. So pure saturated liquid would have a quality of 0%, and pure saturated vapor would have a quality of 100%.

The **vapor pressure** is a measure of the ability of molecules to escape from the surface of a liquid. The lower the vapor pressure, the more the molecules want to stay in the liquid state. Intermolecular attractive forces hold a solid or liquid together. However, the molecules have kinetic energy, which is proportional to the absolute temperature (in Kelvin or Rankine). The attractive forces are trying to keep the molecules in the solid or liquid, in competition with the kinetic energy that is trying to take them out. As more and more molecules accumulate in the space above the liquid more of them will collide with the surface, and some of them will stick to the liquid surface. At a fixed temperature the system will reach a state of dynamic equilibrium, where just as many molecules escape from the surface as accumulate on it. At this point the pressure exerted by the molecules in the space above the solid or liquid is by definition the vapor pressure of the solid or liquid. As the temperature is increased the average kinetic energy increases so more molecules are able to overcome the intermolecular forces and the vapor pressure increases.

We can model vaporization (or condensation) as a constant temperature, constant pressure process, in which the volume and internal energy change. To accomplish such a process there must be heat transfer with the surroundings. The energy to change phase from saturated liquid to saturated vapor is called the latent heat or the heat of vaporization. Various symbols used for this quantity include h_{fg}, L, and Δh.

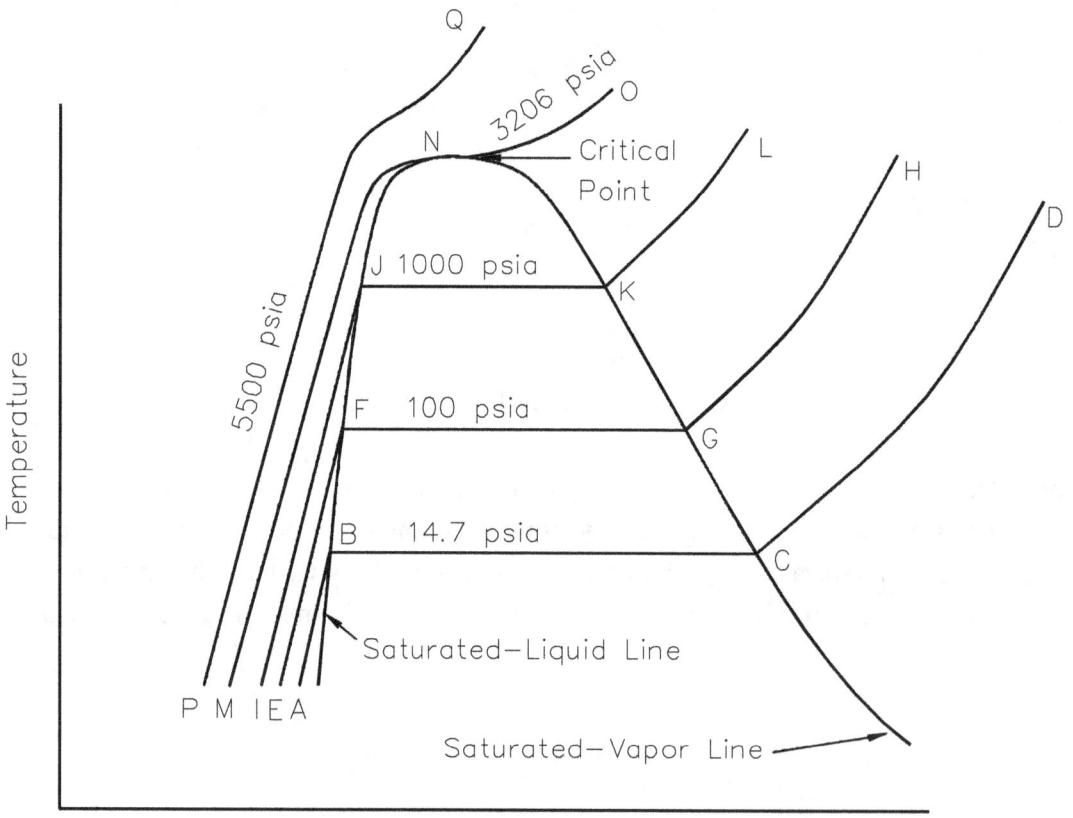

Figure 11: T-v diagram showing states of matter. From [DOE92]

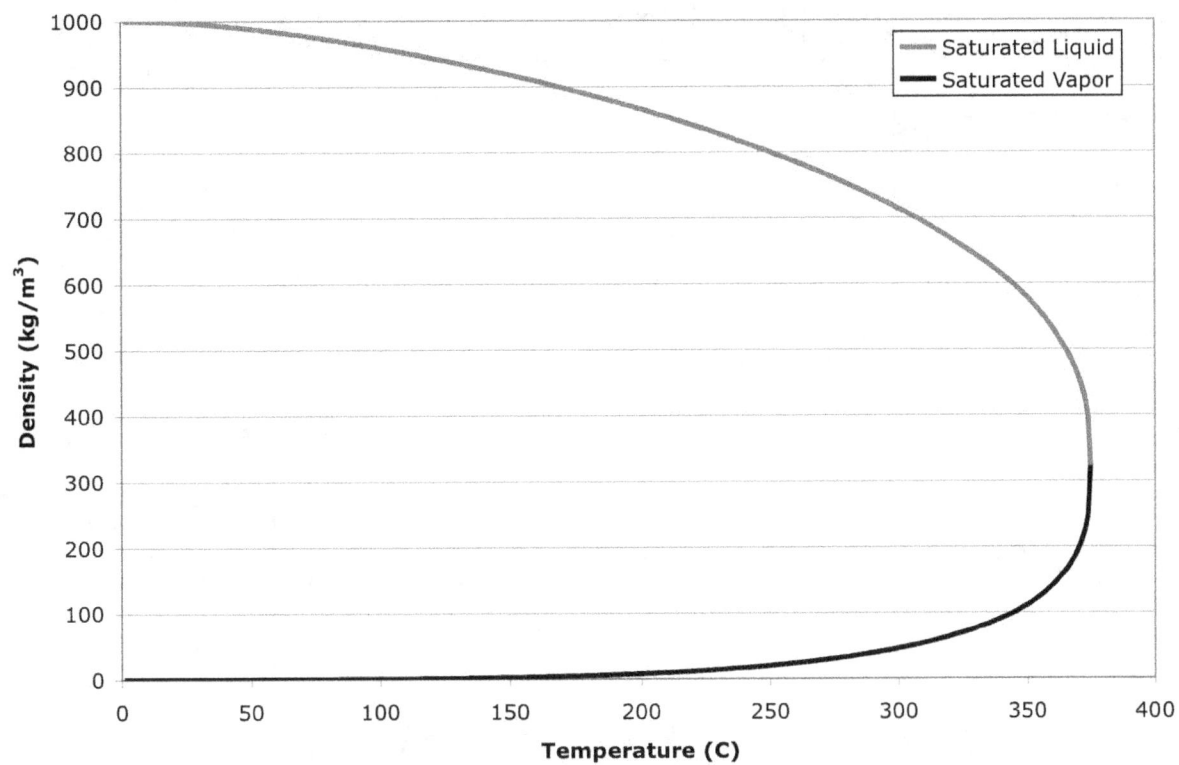

Figure 12: Density-temperature diagram for saturated water.

Your instructor may also make T-s T-h or P-h diagrams but I don't find them useful to solve problems.

The **heat of vaporization** is the amount of energy required to vaporize a substance at a given temperature. The *heat of vaporization* is also called latent heat (L) or enthalpy of vaporization, usually denoted as h_{fg}, but sometimes L. It is a function of temperature (T). A good correlation equation for h_{fg} vs T is:

$$h_{fg}(T) = h_{fg}(T_{bn}) \left(\frac{T_c - T}{T_c - T_{bn}} \right)^{0.38}$$

Figure 13 shows h_{fg} for hydrogen as a function of temperature.

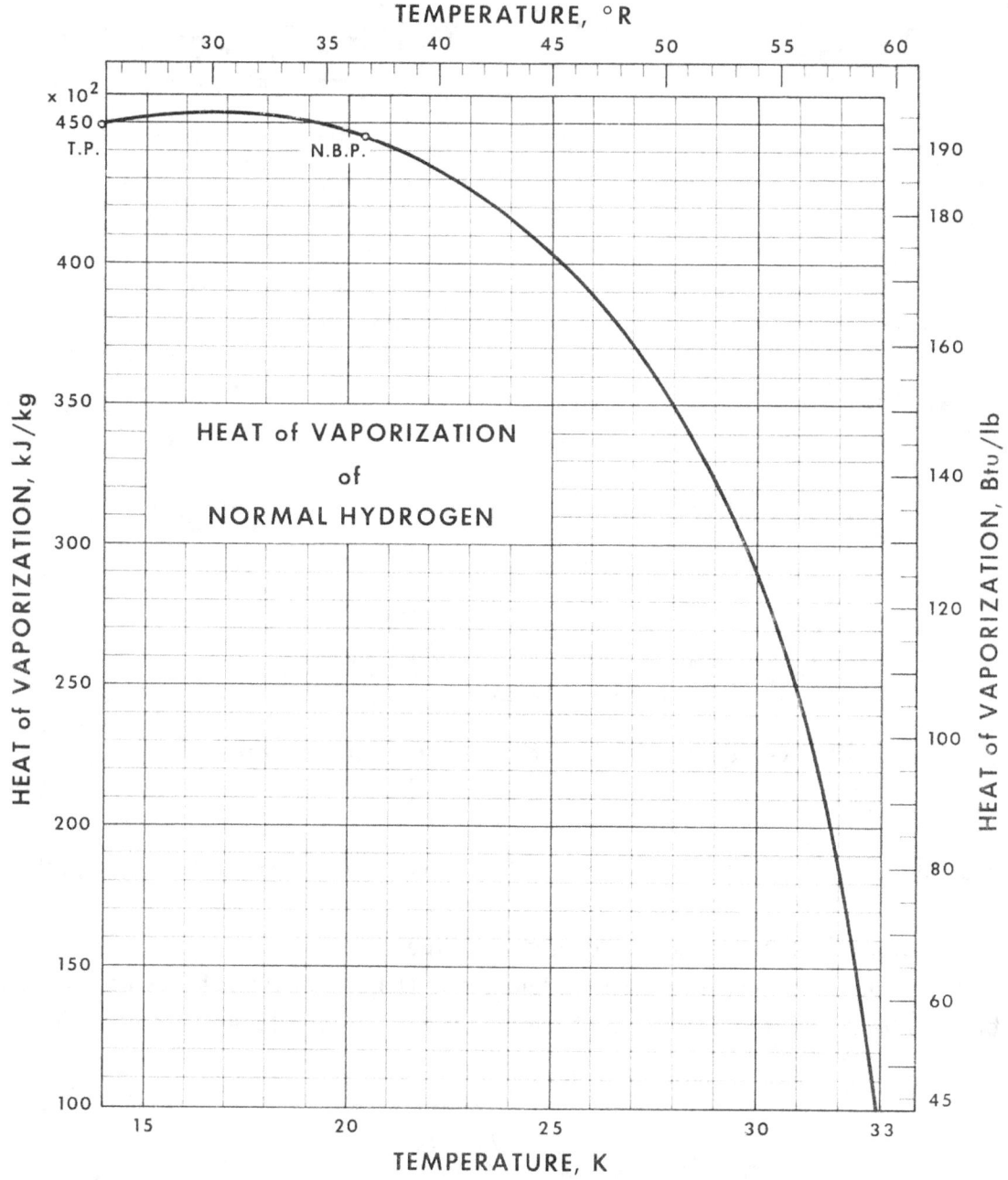

Figure 13: Heat of vaporization as a function of temperature for hydrogen. [NBS81]

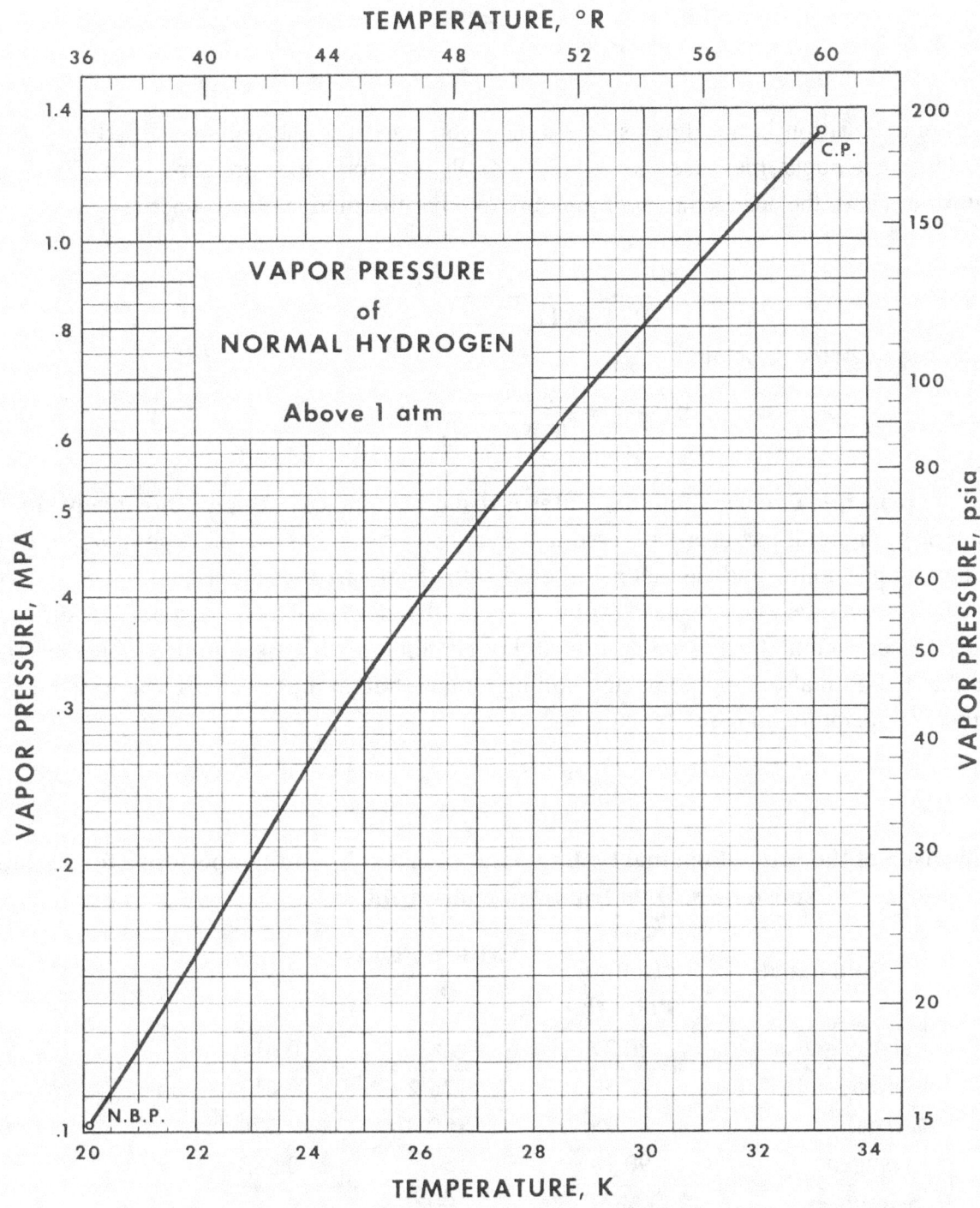

Figure 14: Vapor pressure relationship to temperature for hydrogen. [NBS81]

Curve fits to vapor pressure vs. temperature curves commonly used include the Antoine equation, which is an empirical equation and the values of the constants (A,B,C) must be found for the substance of interest.

$$\log P = A - \frac{B}{C+T}$$

Note that the Antoine equation is not dimensionless, and you must use the units specified to get the correct answer. Different books may use either $P_{sat}(T)$ or $P_{vap}(T)$ for vapor pressure. The **Clausius–Clapeyron equation** relates the change in vapor pressure to other thermodynamic properties:

$$\frac{dP_{sat}}{dT} = \frac{h_{fg}}{T(v_g - v_f)}$$

This equation can be integrated to obtain:

$$\ln\left(\frac{P_1}{P_2}\right) = \frac{h_{fg}}{R}\left(\frac{1}{T_2} - \frac{1}{T_1}\right)$$

The idea behind the **principle of corresponding states** is that all substances exhibit qualitatively similar P-v-T behavior. For example, it has been observed that every substance has a critical point and a triple point, and will have solid, liquid, and gas states. If non-dimensional variables are used, such as reduced temperature, reduced pressure, and reduced volume, then the scaled P-v-T behavior of different substances is found to be quantitatively similar as well. For example, as discussed in [Hirschfelder54], it has been found that the normal boiling point (the boiling temperature at a pressure of 1 atm) is usually about 2/3 of critical temperature, in absolute units:

$$T_b \approx \frac{2}{3} T_c$$

Table 5: Comparison of the ratio of normal boiling point to critical point temperature for various substances. Data from Appendix Table A1.

Substance	Normal Boiling Point (K)	Critical Point (K)	T_b/T_c
Hydrogen	20.27	33.2	0.611
Nitrogen	77.36	126.2	0.613
Argon	87.28	150.9	0.578
Oxygen	90.18	154.6	0.583
Methane	111.70	190.6	0.586
Xenon	165.03	289.7	0.570
Carbon Dioxide	194.75	304.1	0.640
Propane	231.08	369.8	0.625
Ammonia	239.68	406.1	0.590
Water	372.80	647.1	0.576

Using the data in Appendix Table A1, Table 5 shows the ratio of normal boiling point to the critical point for several simple substances. The ratio of T_b/T_c for the selected substances is between 0.57 and 0.64. For the normal alkanes, the ratio of T_b/T_c increases with increasing molecular weight, with n-octane having a value of 0.70.

Gibbs Phase Rule

As stated at the beginning of this section, for a pure substance in equilibrium, two intensive properties are required to define the state (state principle). How many properties (or other pieces of information) are required to fix the state of a multi-component system? For example, a binary water-air mixture that can have both liquid and gas phases – how many pieces of information are needed to fix the state as a unique state?

At equilibrium, both phases will be at the same pressure and temperature. Relative masses can be quantified by mole fractions, and we can use the constraint that $\Sigma\, x_i = 1$. This applies to the overall mixture and each phase. Thus the total number of properties required is:

$$F = C + 2 - P$$

(Notation is **F** properties required to fix the state of **C** components in **P** phases).

So for a pure component, $C = 1$, $P = 1$, and $F = 2$, which is the standard state principle. For a water-air mixture solely in the vapor state, $C = 2$, $P = 1$, and $F = 3$, so in addition to P and T, the relative humidity ϕ (or absolute humidity ω) is required to fix the state.

Also note for a single component (say water) if all 3 phases are present (ice, water, steam) then you are at the triple point, which is a unique point, and $F = 0$. F is also called the number of degrees of freedom of the system. For a substance at the critical point, three phases are present (gas, liquid, supercritical fluid) and no properties are needed to define the state. For a mixture of air and water vapor in the gaseous state, three properties are needed to define the state. The most commonly used ones are pressure, temperature, and the relative humidity. For a pure substance a single-phase region, say in the gas phase, two independent intensive properties are needed to define the state. For a pure substance in the two-phase vapor-liquid region, only one property is needed to define the state. Since there is a unique relationship $P_{sat} = P_{sat}(T)$, if either pressure or temperature is specified than the other is also known.

An equation of state (EOS) is a mathematical expression that describes a system at thermodynamic equilibrium. These equations are empirical relationships that have been developed based on experimental observations. The State Principle states that for a pure substance, it is necessary to define the values of only two properties to completely define the state and set the values of all the other thermodynamic properties. In practice, pressure, temperature, and specific volume (or its reciprocal, density) are the easiest properties to measure

Example 4:
If a closed vessel contains a mixture of two components, and both liquid and gas phases exist in equilibrium with each other, how many properties are needed to define the state?

SOLUTION: We could set up the problem as initial having 6 unknowns: Pressure, temperature, the percentage of substance A in the liquid phase, the percentage of substance B in the liquid phase, the percentage of A in the gas phase, and the percentage of B in the gas phase. We have four constraints on the system: The mole fractions of the two components in the gas phase must add up to 1.0, the mole fractions of the two components in the liquid phase must add up to 1.0, substance A is in equilibrium between gas and liquid phases, and substance B is in equilibrium between gas and liquid phases. So if we have 6 unknowns and 4 constraints, this means that 2 more pieces of information are needed to solve the system. Similarly, according to the Gibbs Phase Rule
$$F = C + 2 - P = 2 + 2 - 2 = 2$$
Confirming that 2 properties are needed to define the state.

Examples of multi-component mixtures include:
- Gas-Gas - such as air
- Liquid-Liquid – emulsions such as milk
- Gas-Liquid – Sprays, mist, fog, carbonated beverages
- Solid-liquid slurries and gels, solid-gas foams (or liquid-gas foams)

Ideal Gas Law

For substances that are not ideal gases (liquids, supercritical fluids, non-ideal gases, mixtures) there are other equations we can use, but also very commonly we use tabular data and interpolation to find the properties we need.

The ideal gas law can be written as:

$$PV = NRT$$

where P is the pressure, V is the volume, N is the number of moles of gas present, R is the university gas constant, and T is the temperature. Using units of kmol, where 1 kmol = 1000 mol, the universal gas constant has a value R = 8314 J/kmol-K. By defining the molar specific volume, $v = V/N$, the ideal gas law can be written as:

$$P = \frac{NRT}{V} = \frac{RT}{v}$$

Since practicing engineers more commonly work in units of mass rather than of moles, the ideal gas law can also be expressed in terms of the mass density, $\rho = m/V$:

$$\rho = \frac{PM}{RT}$$

where M is the molecular weight. The mass, m, is related to the number of moles, N, through

$$m = N*M$$

When gram moles are used, as is often preferred by chemists, the molecular weight is in units of grams per gram-mole (g/mol). When units of kilogram moles are used, which is usually more practical for engineers, the molecular weight is in units of kilograms per kilogram-mole (kg/kmol). When using the ideal gas law, it is always a good idea to convert all the variables to fundamental units (and pressure and temperature to absolute units). In metric, this is Pa for P and K for T, m^3 for V, and kg/kmol for M.

The two main assumptions invoked in deriving the ideal gas law are:
1. The molecules of the gas are point particles
2. The electronic interactions between molecules can be ignored, with perfectly elastic collisions occurring between pairs of molecules

These assumptions are satisfied when the density of the gas is low, in the limit as $\rho \rightarrow 0$. This is achieved when:
1. The pressure (P) is low
2. The temperature (T) is high

A rough rule of thumb is that for noble gases and simple diatomic and triatomic gases at 1 atm pressure and room temperature (300 K) and higher temperatures, the ideal gas law will work to the accuracy you need.

Example 5:
Honda has a natural gas model of the Civic, which has a fuel tank pressurized to 3600 psig. How large would the fuel tank have to be to provide the same range as a Civic with an 8-gallon gasoline tank? Assume both the natural gas and gasoline engines have the same efficiencies, natural gas can be represented as methane as an ideal gas, and the heating value of natural gas is LHV = 47,000 kJ/kg, and the heating value of gasoline is LHV = 43,000 kJ/kg, and gasoline has a density of 750 kg/m^3. (1 Gal =

3.78 L). Heating values will be defined later in the Combustion section of this book. It is the chemical energy released when a fuel is burned in the presence of an oxidizer (usually air).

Note: **psig** is pounds per square inch *gage* pressure, which is the pressure relative to the surrounding atmosphere, as would be measured with a tire pressure gage. Most thermodynamic properties are tabulated in absolute pressure, where 0 **psia** (psi *absolute*) corresponds to a perfect vacuum. P = 0 psig corresponds to P = 1 atm of absolute pressure.

SOLUTION: We can use the ideal gas law to calculate the density of methane. P = 3600 psig = 3615 psia = 24,900 kPa

$$\rho = \frac{PM}{RT} = \frac{(24,900,000\ Pa)(16\ kg/kmol)}{(8314\ kJ/kmol \cdot K)(295\ K)} = 162\ \frac{kg}{m^3}$$

The energy comparison is:

$$E_{gas} = \rho_{gas} V_{gas} LHV_{gas} = \rho_{CNG} V_{CNG} LHV_{CNG} = E_{CNG}$$

$$V_{CNG} = V_{gas} \frac{\rho_{gas} LHV_{gas}}{\rho_{CNG} LHV_{CNG}} = (8\ gal)\frac{\left(750\ \frac{kg}{m^3}\right)(43{,}000\ kJ/kg)}{\left(162\ \frac{kg}{m^3}\right)(47{,}000\ kJ/kg)} = 34\ gal = 129\ L$$

The energy contained in 8 gallons (30.2 L) of gasoline is equivalent to 34 gallons (129 L) of compressed natural gas (CNG). As a result, the CNG version of the Civic has very little trunk space, as the fuel tank has to be expanded to give the car a reasonable driving range. For comparison, liquefied natural gas (LNG) would have a density of around 430 kg/m³, which would result in a fuel tank size of 13 gallons (48 L), which is much smaller, but requires the added complexity of keeping the LNG cold below -162 °C.

Note that the Ideal Gas Law must always be used with **Absolute Temperature.** When the temperature is measured in Kelvin (K) or Rankine (°R), it is referenced to 0 at the state of absolute zero. K = °C + 273.15 and °R = °F + 459.67.

The number of molecules, n, is related to the number of moles, N, through Avagadro's number:

$$N = \frac{n}{N_{AV}}$$

The *compressibility factor*, Z, is defined as:

$$Z = \frac{PV}{NRT} = \frac{Pv}{RT}$$

Z is a measure of the deviation of actual gas behavior from that predicted by the ideal gas law. When Z is close to 1.0 the ideal gas law is valid.

Example 6:
Use the ideal gas law to calculate the density (in kg/m^3) of air at 500 K and 10 atm.

SOLUTION: For engineering purposes, it is usually acceptable to approximate the molecular weight of dry air as M = 29 kg/kmol. 1 atm = 101.3 kPa to four significant digits, so 10 atm = 1.013 × 10^6 Pa = 1.013 MPa. The density can then be calculated as:

$$\rho = \frac{PM}{RT} = \frac{(1{,}013{,}000\ Pa)(29\ kg/kmol)}{(8314\ kJ/kmol \cdot K)(500\ K)} = 7.07\ \frac{kg}{m^3}$$

The tabulated value (Appendix #) is 7.035 kg/m^3, for an error of 0.5%, which is sufficiently accurate for most engineering applications.

Example 7:
Use the ideal gas law to calculate the density (in kg/m^3) of steam at STP and at 1.0 MPa (absolute) and 200 °C.

SOLUTION: For engineering purposes, it is usually acceptable to approximate the molecular weight of pure steam as M = 18 kg/kmol. The conversion from temperature in Celsius to Kelvin is T[K] = T[°C] + 273.15, but since most engineering measurements in practice have accuracies of order +/- 1%, there is usually no need to carry more than 3 significant digits, and it is sufficiently accurate to use T[K] = T[°C] + 273. So for this example the absolute temperature is T = 200 + 273 = 473 K. The density can then be calculated as:

$$\rho = \frac{PM}{RT} = \frac{(1{,}000{,}000\ Pa)(18\ kg/kmol)}{(8314\ kJ/kmol \cdot K)(473\ K)} = 4.58\ \frac{kg}{m^3}$$

The tabulated value (Appendix Table A6 – note 10 bar = 1.0 MPa) is 4.85 kg/m^3, for an error of 5.6%. This is probably not sufficiently accurate. The ideal gas law does not work well for steam because water is a polar molecule, with negative charge accumulated near the oxygen end of the molecule and positive charges around the hydrogen atoms, and so the electronic interactions between water molecules are not negligible.

Steam is one gas that the ideal gas law does not do a good job of modeling. The ideal gas law works well for simple non-polar molecules that are not near the vapor dome and are at high temperature and/or low pressure. The ideal gas law is valid at low pressures and high temperatures. For air and similar gases atmospheric pressure can be considered low pressure, and temperatures at room temperature or above can be considered high temperature. The ideal gas law usually works well for air, oxygen, nitrogen, hydrogen, carbon dioxide, and the noble gases, for example.

For gases for which the ideal gas law is not a good model, there are other equations of state (EOS) correlations that can be used. These include the Redlich-Kwong-Soave and Peng-Robinson equations of state. Both of these have curve-fit parameters that must be found for the specific gas you are using. For example, the Redlich-Kwong-Soave equation is:

$$P = \frac{RT}{v-b} - \frac{a\alpha}{v(v+b)}$$

where the parameters a, b, and α must be found or calculated for the substance of interest. Compare this to the ideal gas law, where $P = RT/v$.

IDEAL GAS MIXTURES

Dalton's Law of Partial Pressures:

$$P = \sum P_i$$

For an ideal gas:

$$P_i = \frac{N_i RT}{V}$$

The total number of moles, N, is:

$$N = \sum N_i$$

And the total pressure is:

$$P = \frac{NRT}{V}$$

Combining these we can write:

$$P_i = x_i P$$

where x is the mole fraction

$$x_i = \frac{N_i}{N}$$

$$\sum x_i = 1$$

In words, each component of an ideal gas mixture behaves as if it alone existed at the temperature and volume of the mixture (or at the temperature and partial pressure). For an ideal gas mixture, the mixture properties can be simply calculated:

$$u = \sum x_i u_i$$

$$h = \sum x_i h_i$$

$$s = \sum x_i s_i$$

(All properties evaluated at the temperature and volume of the mixture). Also stated as, there is no change in component properties due to mixing. We can also define mixture-averaged specific heats:

$$c_{V,mix} = \sum x_i c_{V,i}$$

$$c_{P,mix} = \sum x_i c_{P,i}$$

We can define the volume change due to mixing as:

$$\Delta v = v_{mix} - \sum_i x_i v_i$$

For an ideal gas mixture $\Delta v = 0$, for real gases $\Delta v \neq 0$. Similarly, the heat or enthalpy of mixing is defined as:

$$\Delta h = h_{mix} - \sum_i x_i h_i$$

for an ideal gas mixture, $\Delta h = 0$, and there is no temperature change due to mixing (note this is also assuming no chemical reactions).

The **entropy of mixing**, however, is always > 0, even for idea gases. This is because when we bring any two pure gases together, they will naturally mix due to diffusion and form a mixture without any added energy, but we cannot reverse the process and separate out the two mixed gases without adding external work. For a two-component mixture, the entropy of mixing can be calculated as:

$$\Delta S_{mix} = -NR(x_1 \ln x_1 + x_2 \ln x_2)$$

If multiply this enthalpy of mixing by the temperature, we can get an estimate for the minimum amount of work that would be needed to separate this mixture back into the original pure components. For example if you wanted to remove carbon dioxide (CO_2) from atmospheric air, the minimum energy required to do so could be calculated.

It is also useful to define the **mass fraction**, y, for some applications:

$$y_i = \frac{m_i}{m}$$

where the species mass is:

$$m_i = M \times N_i$$

The mixture-averaged molecular weight (kg/kmol) is computed as:

$$M_{mix} = \sum x_i M_i$$

Thus the mole and mass fractions are related by:

$$y_i = x_i \frac{M}{M_{mix}}$$

Example 8:
A mixture is measured to be 12% CO_2, 0.5% CO, 3.5% O_2, 70% N_2, and 14% H_2O by volume. What is the average molecular weight of the mixture? What is the mass fraction of each component? What is the density of the mixture at standard conditions? How does this compare to air?

SOLUTION: The average molecular weight is:

$$M_{mix} = 0.12(44) + 0.005(28) + 0.034(32) + 0.70(28) + 0.14(28) = 28.7 \frac{kg}{kmol}$$

The mass fractions are as follows:

Table 6: Mole and mass fractions for Example #8.

Gas	CO_2	CO	O_2	N_2	H_2O
Mole fraction	0.120	0.0050	0.035	0.700	0.140
Mass fraction	0.184	0.0049	0.039	0.684	0.088

Assuming an ideal gas mixture, the density can be found as:

$$\rho = \frac{PM_{mix}}{RT} = \frac{(101{,}300\ Pa)(28.7\ kg/kmol)}{(8314\ kJ/kmol \cdot K)(298\ K)} = 1.17 \frac{kg}{m^3}$$

For air at the same conditions:

$$\rho = \frac{PM}{RT} = \frac{(101{,}300\ Pa)(29\ kg/kmol)}{(8314\ kJ/kmol \cdot K)(298\ K)} = 1.19\frac{kg}{m^3}$$

These values are close, which is why air is often used as an approximation for exhaust gas properties.

Simple Devices

The simple devices discussed in this section use a thermodynamic **process**, or a sequence of processes to accomplish a task. We will start with piston-cylinder devices.

Piston-Cylinder

Engines are the primary example of this device. Figure 15 shows an illustration for a typical 4-stroke gasoline-fueled engine. Your thermo book probably calls this a "closed system", in that once the valves are closed, the mass of air does not change, but is fixed. If the cylinder compresses (or expands) quickly enough, no heat will be transferred to the surroundings, and the process will be **adiabatic**.

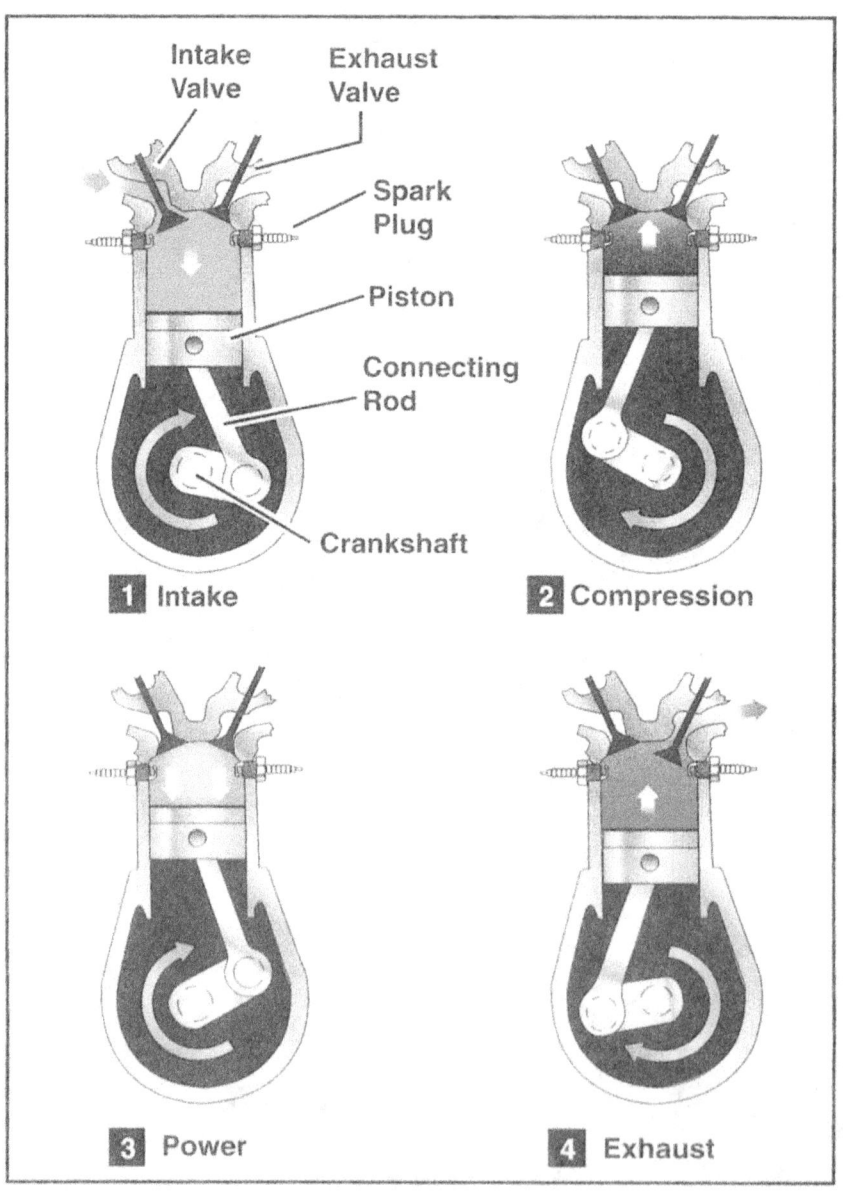

Figure 15: Pictures of the strokes in a typical 4-stroke engine, from [FAA]

So let's consider an engine with a compression ratio of 13:1 (this means the final volume when the gases in the cylinder are fully compressed is 1/13 of the initial volume in the cylinder before compression. To make the math easy, let's assume the pressure of the air in the cylinder when the valves close is 1.0 atm. (In reality the pressure will be less than the surrounding atmospheric pressure, unless you have a turbocharged engine). Let's also take the initial volume of the cylinder as 2.0 L. Let us also assume the initial temperature of the air in the cylinder is 26.85 °C = 300 K. (You quickly learn when working ideal gas problems to put the temperature in absolute units of Kelvin rather than Celsius, or you will get the wrong answer. If your course instructor is making you do these problems in English units, I am sorry for you.)

So when the piston moves to its final position, the volume of air in the cylinder will be:
$$2.0/13 = 0.154 \text{ L} \quad (\text{note } 1 \text{ m}^3 = 1000 \text{ L})$$
So now we need to figure out the final pressure and temperature in the cylinder. If we model the compression process as **isothermal** (the temperature does not change) then the calculation is easy, as the final temperature, $T_2 = T_1 = 300$ K. Then we can use the ideal gas law to figure out the final pressure. Unfortunately in real life such processes are not isothermal. In fact, when you compress a gas, it gets hotter. (The main reason I included the isothermal calculation is because your instructor may ask it on an exam – I do not know of a practical example where it occurs).

So we have to figure out how much hotter the gas gets. Now normally at this point one of those expensive textbooks would go into a long explanation of how we figure out exactly how much hotter the gas gets, and your instructor may feel obligated to spend a good deal of class time going over that. I however am just trying to sell books, so I am going to cut straight to the answer.

We have found from experience over the years that we can **model** the compression of a piston-cylinder device that moves reasonably quickly (as the pistons in your car engine do) as an **isentropic** process. You may get asked to define the term isentropic on a quiz, so I should tell you the word means constant entropy, in this case $s_2 = s_1$. (Recall we introduced entropy with the 2nd law when talking about waste heat – an isentropic process is also an adiabatic process, so no heat is lost from our gas). For the special case of an ideal gas, we find the relationship:

$$P_1 V_1^k = P_2 V_2^k$$

Where k is the ratio of specific heats. (some textbooks also use the Greek letter γ for k).

$$k = \frac{c_p}{c_v}$$

For air at standard atmospheric conditions, k = (1.005 kJ/kg-K) / (0.718 kJ/kg-K) = 1.4.

Thus for our example problem, $P_2 = P_1 (V_1/V_2)^k = (1.0 \text{ atm})(13)^{1.4} = 36.3$ atm.

We can also figure out the temperature at the end of compression using the ideal gas law with a constant number of moles:

$$\frac{P_1 V_1}{T_1} = \frac{P_2 V_2}{T_2}$$

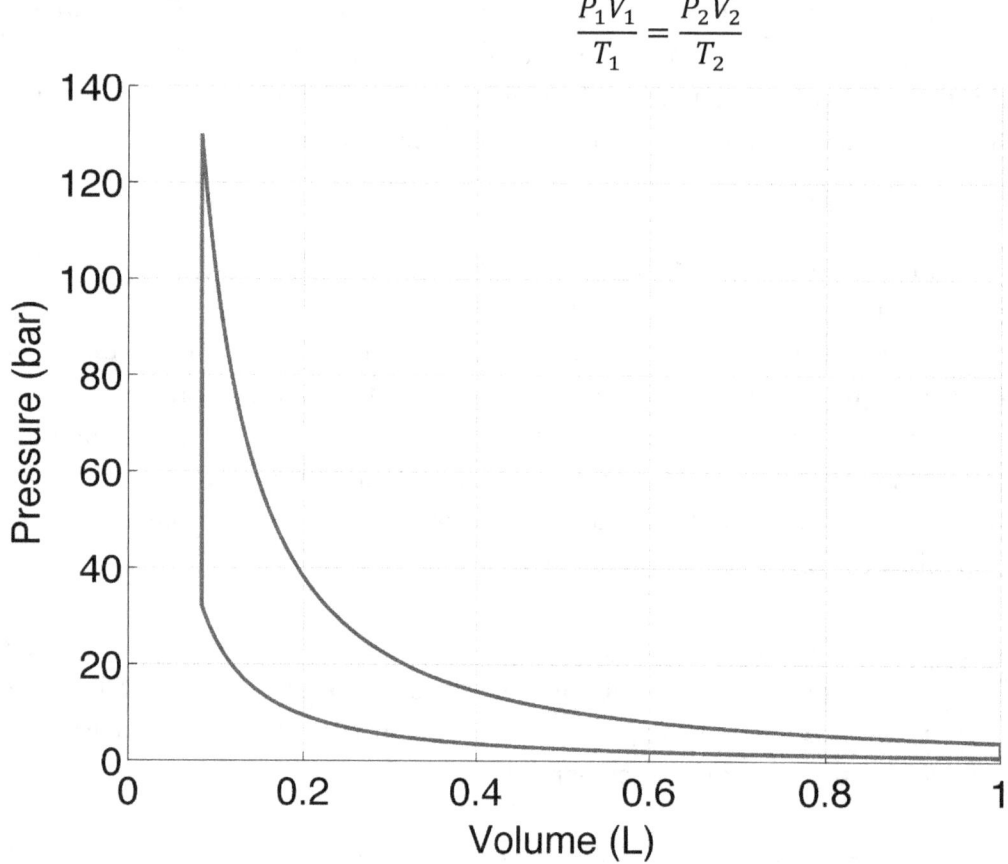

Figure 16: Pressure-Volume (P-V) diagram of the Otto cycle, for conditions of Example #9, with a compression ratio of 12 and initial volume of 1.0 L.

Table 7: Steps in the Otto Cycle

Process	Thermodynamic relations	Volume relations
Isentropic Compression	$P_1 V_1^k = P_2 V_2^k$	$V_2 = V_1$ / Compression ratio
Constant Volume Heat Addition	$Q_{IN} = m\, c_V\, (T_3 - T_2)$	$V_3 = V_2$
Isentropic Expansion	$P_4 V_4^k = P_3 V_3^k$	$V_4 = V_3$ * Compression ratio
Constant Volume Heat Rejection	$Q_{OUT} = m\, c_V\, (T_4 - T_1)$	$V_4 = V_1$

Example 9:
Calculate the states of the cold-air-standard Otto Cycle for a compression ratio of 12, with intake conditions at P = 1.0 bar and T = 300 K, assuming wide-open throttle. The heat release from combustion is Q/m = 1750 kJ/kg and the maximum volume in the cylinder is 1.0 L. What is the efficiency of the cycle?

SOLUTION: The initial state properties are given in the problem statement. To find the properties at the end of the compression stroke, the relationship for an isentropic compression of an ideal gas with constant specific heats can be used:

$$PV^k = C$$

Thus $P_1V_1^k = P_2V_2^k$, and the pressure at state 2 can be found from:

$$P_2 = P_1 \left(\frac{V_1}{V_2}\right)^k = P_1(CR)^k = (1.0 \, bar)(12)^{1.4} = 32.4 \, bar$$

The temperature at state 2 can be found using the ideal gas law.

$$\frac{P_1V_1}{T_1} = \frac{P_2V_2}{T_2}$$

For a fixed mass of gas, combining the ideal gas law with the isentropic compression relation, $P_1V_1^k = P_2V_2^k$, yields:

$$T_2 = T_1(CR)^{k-1} = (300 \, K)(12)^{0.4} = 811 \, K$$

When using the ideal gas law, the temperature and pressure must always be expressed in absolute units. For constant volume combustion with constant specific heats, $Q_{chem} = m \, c_V \, \Delta T$.

$$\Delta T = \frac{Q}{m}\frac{1}{c_V} = \frac{1750 \, kJ/kg}{0.718 \, kJ/kg \cdot K} = 2437 \, K$$

Thus the temperature at the end of combustion is $T_3 = T_2 + \Delta T = 811 \, K + 2437 \, K = 3248 \, K$. Once again the ideal gas law can be used, this time to find the pressure. Since the combustion process occurs at constant volume, then

$$\frac{P_3}{T_3} = \frac{P_2}{T_2}$$

$$P_3 = (32.4\ bar)\frac{3248\ K}{811\ K} = 129.8\ bar$$

For the isentropic expansion of the power stroke the pressure-volume relationship is:

$$P_4 = P_3\left(\frac{V_3}{V_4}\right)^k = P_3\left(\frac{1}{CR}\right)^k = (129.8\ bar)\left(\frac{1}{12}\right)^{1.4} = 4.0\ bar$$

The final temperature is:

$$T_4 = T_3\left(\frac{1}{CR}\right)^{k-1} = (3248\ K)\left(\frac{1}{12}\right)^{0.4} = 1202\ K$$

The efficiency of the Otto cycle can be calculated from an energy balance:

$$\eta = \frac{Q_{IN} - Q_{OUT}}{Q_{IN}}$$

where $Q_{OUT} = c_V(T_4-T_1) = 0.718$ kJ/kg-K (1202 K – 300 K) = 648 kJ/kg.

$$\eta = \frac{1750\ kJ/kg - 648\ kJ/kg}{1750\ kJ/kg} = 0.630 = 63.0\%$$

Table 8: Results for Otto Cycle Example

State	Volume (L)	Pressure (bar)	Temperature (K)
1	1.0	1.0	300
2	0.083	32.4	811
3	0.083	129.8	3248
4	1.0	4.0	1202

If you are interested in learning more about engines, you may want to check out my book titled *Thermodynamics of I.C. Engines* (shameless plug warning). Howstuffworks.com is also a good place to get basic explanations of various mechanical devices, including car and diesel engines.

<u>Water pumps</u>

Pumps are devices used to raise the pressure of a liquid. Historically pumps were often used to raise the elevation (gravitational potential energy) of water. Liquids have the convenience that their density does not change significantly for most practical conditions. Thus we can model liquids as **incompressible** fluids (no matter how much you increase the pressure, the volume will not decrease). The density of a liquid will change slightly with changes in temperature, but almost not at all with pressure.

A general form of the first law of thermodynamics for **open systems** is:

$$\frac{dE}{dt} = \sum \pm \dot{Q} + \sum \pm \dot{W} + \sum_{in} \dot{m}\left(h + \frac{V^2}{2} + gz\right) - \sum_{out} \dot{m}\left(h + \frac{V^2}{2} + gz\right)$$

In my experience we can almost always through away the potential and kinetic energy terms when analyzing thermodynamic devices. We also consider a pump that operates in steady state (dE/dt = 0) with no significant heat loss to the environment (Q = 0). This simplifies the first law to:

$$\dot{W} = \dot{m}(h_2 - h_1) = \dot{m}\left[(u_2 - u_1) + \left(\frac{P_2}{\rho} - \frac{P_1}{\rho}\right)\right]$$

Here I have used the definition of enthalpy (h = u + Pv) and that density is the reciprocal of specific volume:

$$\rho = \frac{1}{v}$$

Pressure can be thought of as a form of energy. The "Pv" term in enthalpy (h = u + Pv) is also called the flow work. I prefer density to specific volume since for liquids usually either the density is provided or the specific gravity (density relative to water). For water at standard conditions the density is 1000 kg/m³ = 1 kg/L, and SG = 1.0.

So if the temperature of the liquid does not change as it goes through the pump ($T_2 = T_1$) then $u_2 = u_1$, and the first law simplifies to:

$$\dot{W} = \dot{V}\Delta P$$

where (V) is the volume flow rate. This is often expressed in liters per minute (LPM) or gallons per minute (GPM), which then needs to be converted to cubic meters per second. Note we run out of letters in the alphabet, hence using some letter in the Greek alphabet, but sometimes even then we run out and have to re-use symbols. In some older books Q is used for volume flow rate (m³/s), but this is confusing with the symbol for heat transfer. V is used for velocity, and also for volume. To avoid confusion, often volume is V with a bar through it: V̶.

Hint: A good way to tell if you have a valid equation is to check that units on both sides of the equation match up. If they are the same, you probably have the right terms, but if they are different you know you are wrong and need to start again.

Example 10:
A particular centrifugal pump can provide 250 ft head at 760 gpm through a 10.5" pipe, while consuming 50 kW of electrical energy. Calculate the efficiency of the pump.

SOLUTION: If you are given a pump problem in English units, convert everything to metric. It will be easier, and then you can convert answer back to English units at end if needed. So in metric we have:

Variable	English	Metric
Head	250 ft	76.2 m
Flowrate	760 GPM	2877 L/min
Diameter	10.5 inches	0.267 m
Rated Power	67 hp	50 kW

In pumps, it is common to give their rating in "head" or units of elevation, since pumps are commonly used to pump liquids from a low level to a higher one. We can convert this to gravitational potential energy, in units of pressure, by:

$$P = \rho g h$$

Thus the pressure rise the pump can generate at this point, for a working fluid of water, is:

$$P = \left(1000 \frac{kg}{m^3}\right)\left(9.8 \frac{m}{s^2}\right)(76.2 \, m) = 746{,}760 \, Pa = 746.8 \, kPa$$

There is an old saying in politics, "follow the money". In engineering, we should say, "follow the units." If your units don't work out right, you have probably used the wrong equation. For this example, 1 Pa = 1 N/m^2 = 1 kg/m-s^2, which is the same on both the left and right sides of the equation. Also note that for the accuracy of typical engineering calculations (usually within 1-2% is sufficient) we can use a value of 9.8 m/s^2 for g.

Now we still need to do something about the units on flow rate. LPM is commonly used for water flows, but won't do when we need to calculate the theoretical pump power. So we need to convert LPM to something in base units (combination of kg, m, and s):

$$2877 \frac{L}{min} \times \frac{1 \, m^3}{1000 \, L} \times \frac{1 \, min}{60 \, s} = 0.04795 \frac{m^3}{s}$$

Now we can calculate the mechanical power output of the pump from the first law:

$$\dot{W} = \dot{V} \Delta P = \left(0.04795 \frac{m^3}{s}\right) 746{,}760 \, Pa = 35{,}800 \, W = 35.8 \, kW$$

To calculate the efficiency of the pump, we take the ratio of the mechanical output power to the electrical input power:

$$\eta = \frac{35.8 \, kW}{50.0 \, kW} = 0.72 = 72\%$$

Compressors

When we want to increase the pressure of a gas instead of a liquid, things are more complex than they are in a liquid pump because gases are **compressible**. A compressor can be defined as a mechanical device that draws in ambient air and expels it at a higher pressure. There are many uses for compressed air, including the powering of pneumatic tools, conveyers, pneumatic controls and actuators, spraying coatings, injection molding, pressure treatment, glass blowing, snow making, starting gas turbines, air brakes, and cleaning [DOE03].

Figure 17: Refrigeration compressors. Source NASA Glenn Image Archives.

Practically all industrial plants have compressed air systems, ranging in size from 5 hp to more than 50,000 hp. There are many different types of compressors, which can be classified into two broad categories of positive-displacement and dynamic compressors, as shown in Figure 18.

Centrifugal Air Compressors

The most common centrifugal air compressors produce flow rates from 1200 to 5000 cubic feet per minute (CFM), equivalent to 0.57 to 2.4 m^3/s, but compressors with flow rates varying anywhere from 300 to 100,000 CFM (0.14 to 47 m^3/s) are available. [DOE03]

Sizing Compressors

Compressors are more efficient at rated load than at part load, so it is important to use a properly sized compressor to maximize efficiency. In some cases, it may be preferable to have multiple smaller compressors rather than one larger compressor to maintain high efficiency over variable compressed air demands. [DOE03] Additionally, a receiver tank is usually added to a compressed air system to provide capacitance to the system and minimize cycling of the compressor on and off. The U.S. Dept. of Energy estimates that typically about ¾ of the cost of a compressed air system is for the electricity to run the compressor, with the remainder split between the initial cost to buy the equipment and maintenance. [Tipsheet04]

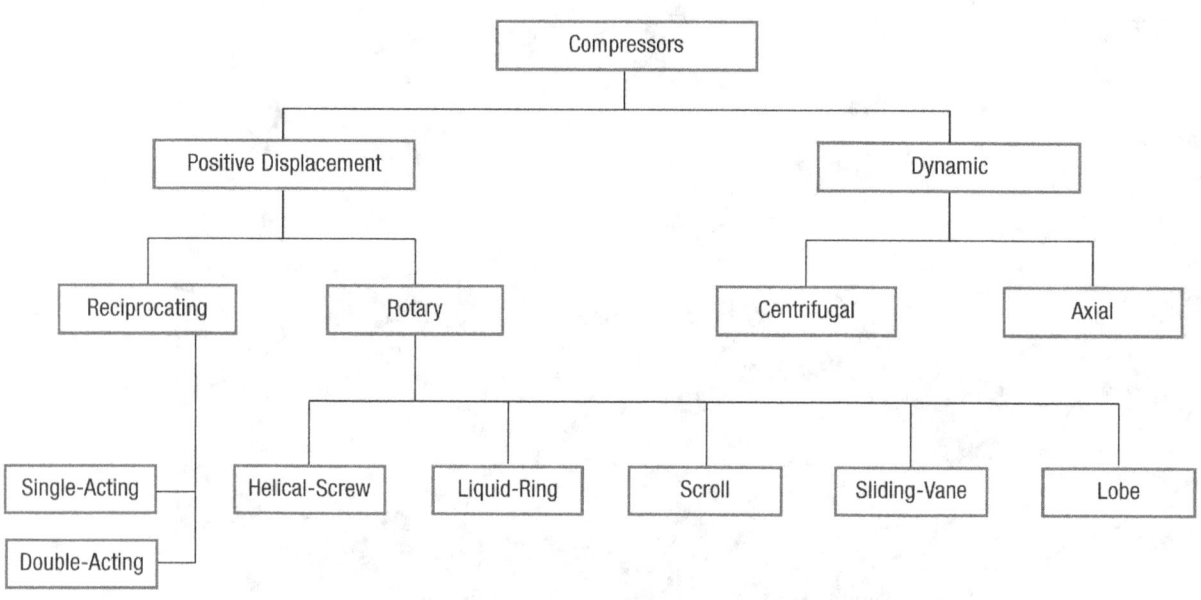

Figure 18: Different types of air compressors, from [DOE03].

For an <u>isothermal compression</u> of an ideal gas, the work of compression for a finite amount of gas can be calculated using the boundary work:

$$W = \int P \, dV$$

The ideal gas law is:

$$P = \frac{NRT}{V}$$

This can be substituted into the boundary work equation:

$$W = \int \frac{NRT}{V} \, dV = NRT \int \frac{1}{V} \, dV$$

Applying limits of integration from state 1 to state 2:

$$W = NRT(\ln V_2 - \ln V_1) = NRT \ln \frac{V_2}{V_1}$$

For an isothermal process of an ideal gas of fixed mass, the volume ratio is inversely proportional to the pressure ratio:

$$W = NRT \ln \frac{P_1}{P_2}$$

Using the ideal gas law, PV=NRT, again.

$$W = P_2 V_2 \ln \frac{P_1}{P_2}$$

The efficiency of a compressor can be defined as the ratio of the amount of work required in an isothermal compression process to the actual work done.

$$\eta_{compressor} = \frac{W_{isothermal}}{W_{actual}}$$

Example 11:
How much work is required to fill a receiving vessel of volume (50 gallons = 189 L) with air at 100 psig, assuming an isothermal compression process?

SOLUTION: The final pressure is P_2 = 100 psig = 114.7 psia = 791 kPa. Assuming the initial pressure to be at a standard atmosphere, P_1 = 14.7 psia = 101,325 Pa = 101.3 kPa, then the volume of atmospheric air that needed to be drawn into the compressor can be calculated as:

$$V_1 = V_2 \frac{P_2}{P_1} = (189\ L)\frac{791\ kPa}{101\ kPa} = 1475\ L$$

The work of compression is calculated using the equation on the previous page:
$$W = P_2 V_2\ ln\frac{P_1}{P_2} = (791{,}000\ Pa)(0.189\ m^3)\ln\left(\frac{101}{791}\right) = -307{,}600\ J = -307.7\ kJ$$

The mass of air compressed is:
$$m = \rho V = \left(1.2\frac{kg}{m^3}\right)(1.475\ m^3) = 1.77\ kg$$

On a per mass basis, the work of compression is -307.7 kJ/1.77 kg = 174 kJ/kg.

What is the difference between a fan, a pump, and a compressor? A fan moves without a significant change in pressure, while a compressor adds pressure to a gas (compressible substance). A pump adds pressure to a liquid (incompressible substance).

Another question is what is the model for an ideal compressor? Normally we think of an isentropic process as the ideal to be obtained in thermodynamics, but isentropic compression of a gas with no heat loss results in the gas heating up, which reduces the density. In comparison, an isothermal compression process, where heat is exchanged to the surroundings to cool the gas being compressed, will raise the density and thus increase the total compression obtained, even though entropy increases.

Figure 19: Fans supplying a wind tunnel. Source NASA Image Archives.

Turbines

A turbine is the reverse of a compressor, in that it extracts energy from a pressurized gas to do mechanical work, unlike a compressor that does mechanical work to add pressure to a gas. Two common types of turbines are steam turbines, which generate power in Rankine-cycle power plants powered by coal or nuclear fission, where water is the working fluid, and gas turbines, which use exhaust gas from a combustion process, to generate power.

Steam Turbines

Steam turbines can vary greatly in size. Siemens sells small steam turbines in the range of 10 MW and less, such as their SST40, that operates at 40 bar and 400 °C and produces from 75-300 kW of power. The 10 MW SST120 uses steam at 131 bar and 530 °C. The Siemens SST900 can handle steam at the inlet of 165 bar and 585 °C, producing up to 250 MW of power.

Also note that real turbines do not behave isentropically. An appropriate isentropic efficiency of the turbine can be defined as:

$$\eta_{turbine} = \frac{\Delta h_{act}}{\Delta h_{isen}} = \frac{h_3 - h_{4,a}}{h_3 - h_{4,s}}$$

Example 12:
Steam at 100 bars and 400 °C enters a turbine and expands to 10 bars and 200 °C. (convenient numbers for looking up values in a table, but not likely so convenient in reality) The ambient conditions are 1 bar and 27 °C. If the turbine is well-insulated, determine:
- the actual work developed, per mass of steam
- the maximum isentropic work that could be developed with an expansion to 10 bars
- the isentropic efficiency of the turbine

SOLUTION: using website: https://www.steamtablesonline.com/Steam97Web.aspx
At P = 100 bar and T = 400 °C, h_1 = 3097.375 kJ/kg and s_1 = 6.213929 kJ/kg-K
At P = 10 bar and T = 200 °C, h_2 = 2828.267 kJ/kg and s_2 = 6.695488 kJ/kg-K
Note: Values in Appendix Table A6 are slightly different due to rounding.

$$\frac{W}{m} = h_1 - h_2 = 3097.375 \frac{kJ}{kg} - 2828.267 \frac{kJ}{kg} = 269.468 \frac{kJ}{kg}$$

To find the efficiency, we have to find a point at 10 bars that has the same entropy, so we have to find the state where P = 10 bar and s = 6.213929 kJ/kg-K. Recall the state principle for a pure substance we need 2 properties to define the state, which we have (P,s). To use steamtablesonline.com with this function requires purchasing an account, but we can also find the data we need using the free database at NIST's site:

https://webbook.nist.gov/chemistry/fluid/

by selecting water and generating isobaric data at P = 10 bar over the temperature range of interest. Then select display data in HTML table. The relevant section is below:

Table 9: Relevant property data for water.

Phase	T (°C)	P (bar)	ρ (kg/m³)	h (kJ/kg)	s (kJ/kg-K)
Liquid	179.88	10	887.13	762.52	2.1381
Gas/Vapor	179.88	10	5.145	2777.1	6.5850

By looking at isobaric data we discover the answer will be a 2-phase mixture at saturation temperature of 179.88 °C. So we need to use linear interpolation to find the quality of the 2-phase mixture that matches the entropy (can use your calculator, EXCEL, MATLAB, whatever you like):

$$x = \frac{s - s_f}{s_g - s_f} = \frac{6.2139 - 2.1381}{6.5850 - 2.1381} = 0.917$$

$$h = (1-x)h_f + (x)h_g = (0.083)(762.52) + (0.917)(2777.1) = 2609.9 \, kJ/kg$$

$$\eta = \frac{h_1 - h_2}{h_1 - h_{2,s}} = \frac{3097.4 - 2828.3}{3097.4 - 2609.9} = 0.552 = 55.2\%$$

Typically the isentropic efficiency of steam turbines varies from 20 to 70%. [EERE04] Multi-stage condensing turbines will be more efficient that single-stage turbines. An interesting steam turbine calculator tool (though unfortunately in British units) can be found at:

https://www4.eere.energy.gov/manufacturing/tech_deployment/amo_steam_tool/equipTurbine

Electric motors and batteries

The electrical power input to a DC motor is:

$$\dot{W} = Vi = (iR + E)i = i^2R + Ei$$

Where V is the voltage in volts, i is current in amps, R is the resistance of the windings in ohms, and E is the back emf (electromotive force), which depends on the speed of the motor. The power lost as heat is i^2R, so the net useful electrical power is Ei. The mechanical output power of the motor is:

$$\dot{W} = T\omega$$

where T is the torque [N-m] and ω is the rotational speed [rad/s]. Rotational speed is commonly measured in revolutions per minute [RPM], so it is necessary to convert between the two units:

$$1 \; RPM = 1\frac{rev}{min} = 1\frac{rev}{min} \times \frac{2\pi \; rad}{1 \; rev} \times \frac{1 \; min}{60 \; s} = \frac{2\pi}{60} \frac{rad}{s}$$

The efficiency of the motor can be defined as:

$$\eta = \frac{\dot{W}_{output}}{\dot{W}_{input}} = \frac{T\omega}{Vi}$$

Motors can be over 90% efficient, as shown in Figure 20. Note in the US motor output will often be given in mechanical horsepower, while input to the motor will be in Amps and Volts (and hence easy to calculate kW). Also note 1 hp = 0.746 kW.

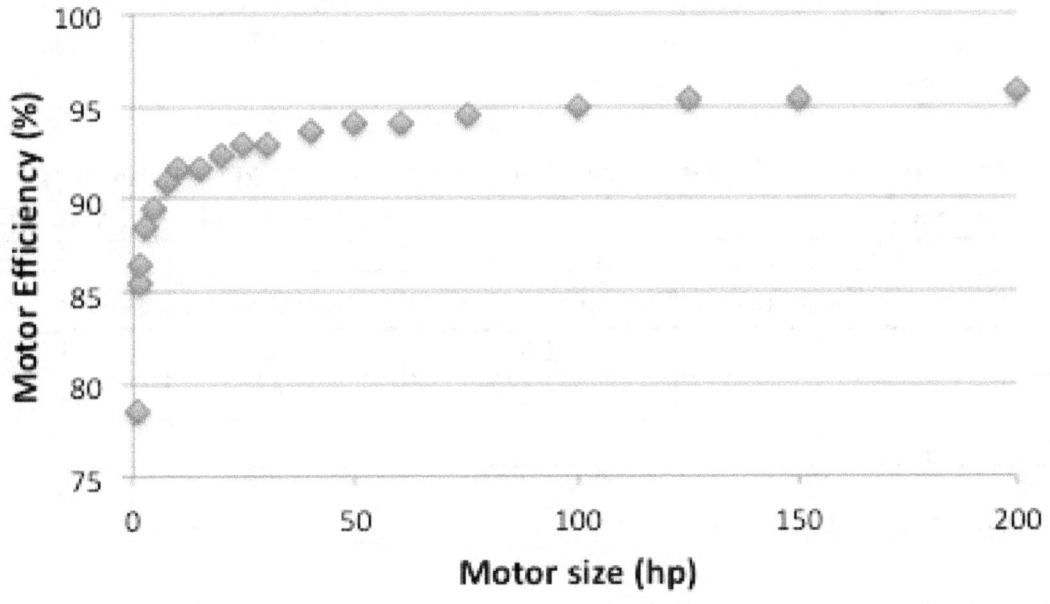

Figure 20: Typical minimum efficiency of electric motors as a function of motor size.

Generators
There is no real difference between at DC motor and a DC generator, except for the direction of power and current flow.

<u>Losses in Motors and Generators:</u>
- Copper loss – the i^2R loss due to resistance of the windings
- Iron (core) loss – Due to hysteresis and eddy current losses in the armature (magnetic effects)
- Mechanical losses – bearing and brush contact friction, air resistance against moving parts

Efficiency

Large three phase motors are more efficient than smaller 3-phase motors, and most all single-phase motors. Large induction motor efficiency can be as high as 95% at full load, though 90% is more common. Efficiency for a lightly load or no-loaded induction motor is poor because most of the current is involved with maintaining magnetizing flux. As the torque load is increased, more current is consumed in generating torque, while current associated with magnetizing remains fixed.

Batteries

Batteries were mentioned briefly in the section on Energy Storage devices. A battery is an electro-chemical energy storage device. A battery contains a fixed quantity of chemical fuel. For many types of batteries the chemical reactions are reversible and the battery can be recharged. Batteries contain two electrodes and an electrolyte solution. Like fuel cells, batteries are not heat engines and so are not limited by the Carnot efficiency limit. Whereas a fuel cell must be continuously supplied to a fuel cell, a battery contains a fixed quantity of chemical fuel. For many types of batteries the chemical reactions are reversible and the battery can be recharged. Batteries contain two electrodes and an electrolyte solution. The lead-acid battery was the first practical storage battery, invented in the 1850s. It uses lead electrodes and an acid electrolyte (aqueous sulfuric acid).

Table 10: Comparison of energy densities of different types of batteries. Most batteries have operating temperature range of -20 to 60 °C. [ANL10]

Type	Energy Storage (W-hr/kg)	Nominal Voltage per Cell (V)	Self Discharge per Month	Overcharge Tolerance
Lead Acid	35-50	2.0	5%	High
NiCd	40-60	1.2	20%	Moderate
NiMH	75-95	1.2	30%	Low
Li-ion	120+	3.6-4.0	10%	Very Low

*Note: 1 Watt-hour = 3600 J = 3.6 kJ, so 1 kW-hour (kWh) = 3600 kJ = 3.6 MJ

The **capacity** of battery is often given in Amp-hours. So if a battery is rated at 50 Ah, it can provide a current of 5 Amps for 10 hours before being completely discharged. The SAE J1772 connector is the standard for re-charging EVs and PHEV in North America. (There is an SAE Standard for pretty much everything relating to vehicles). The J1772 connector is a 5-pin connector 4.3 cm in diameter, allowing for 240 V, 40 A charging. Compare this to a standard 120 V, 10 A household circuit – how much longer would it take to charge the same amount of energy?

Hybrid vehicles supplement a traditional gasoline engine with an electric motor and a small battery pack, so that the engine can operate in a more efficient part of its operating regime, be sized smaller, and that a portion of the kinetic energy of the vehicle can be recovered in regenerative braking, usually only about 40-50% [NREL02]. Table 11 shows a selection of hybrid electric-gasoline cars currently on the market in North America, and Table 12 a corresponding sample of pure electric vehicles.

Table 11: Selected hybrid gasoline(petrol)-electric cars currently on the market (2020).

Car	Engine size (L)	Motor (kW)	Battery Energy (kW-hr)	Range (miles)	Range (km)	MPG	Cost ($USD)
Ford Fusion	2.0	88	1.4	590	950	42	$28,000
Kia Niro	1.6	32	1.56	600	960	50	$23,500
Honda Insight	1.5	96	1.2	550	890	52	$23,000
Toyota Prius	1.8	53	1.3	630	1020	56	$24,000
Toyota Prius Prime (plug-in)	1.8	53	8.8	25 (elec) 640 (total)	40 (elec) 1030	54 133(e)*	$28,000

*e is an average to account for portions of driving in electric-only mode

Table 12: Selected electric cars currently on the market. (2020)

Car	Motor (kW)	Motor (hp)	Battery Energy (kW-hr)	Range (miles)	Range (km)	Cost ($USD)
Ford Focus EV	92	123	33	115	185	$29,000
Hyundai Ioniq	88	118	28	124	200	$30,000
Volkswagon eGolf	100	134	36	125	201	$30,500
Nissan Leaf	110	147	40	151	243	$30,000
BMW i3	137	184	42	153	246	$44,500
Chevy Bolt	149	200	60	259	417	$36,600
Tesla Model 3	250	340	80	310	499	$44,000
Tesla Model S	310	416	100	373	600	$94,000

Multi-component Devices

If we put together combinations of the simple devices discussed in the previous section into a loop, we can build a thermodynamic cycle. You may notice that a lot of the simple thermodynamic cycles have 4 steps (or state points or devices).

Thermal Power Plants

Steam Power Plants

The basic Rankine cycle is commonly used to represent steam power plants. The four steps of the Rankine cycle are isentropic compression from a saturated liquid state, constant pressure heat addition, isentropic expansion, and constant-pressure heat rejection. Figure 22 shows a T-s diagram of the basis Ranking cycle with water as the working fluid.

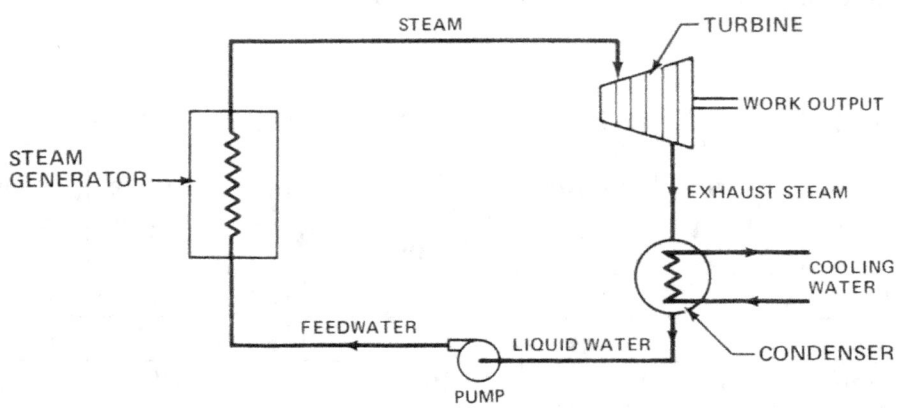

Figure 21: Components of a basic Rankine cycle power plant. [Deskbook]

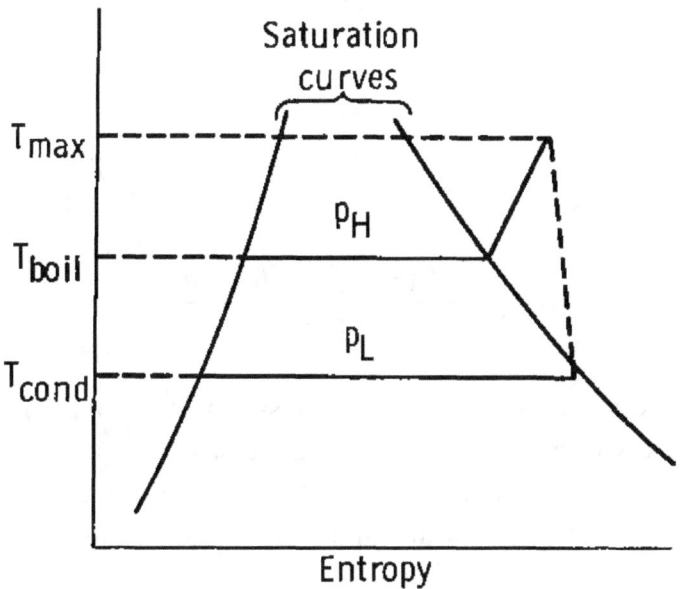

Figure 22: T-s plot of simple Rankine cycle. [NASA69]

Table 13: Steps in the basic Rankine Cycle

Process	Thermodynamic relations
Isentropic Compression of Liquid	$s_2 = s_1$
Constant Pressure Heat Addition	$P_3 = P_2$
Isentropic Expansion	$s_4 = s_3$
Constant Pressure Heat Rejection	$P_1 = P_4$

Note that the steps in a Rankine cycle are the same as the steps in a Brayton cycle. The difference is that the working fluid in a Brayton cycle remains a gas throughout the entire cycle, while phase change takes place in a Rankine cycle.

The efficiency of the simple Rankine Cycle is the net work produced divided by the heat input from combustion.

$$\eta = \frac{\dot{W}_{net}}{\dot{Q}_{in}}$$

The net work is the work produced by the turbine minus the work required to run the pump.

$$\dot{W}_{net} = \dot{m}(h_3 - h_4) - \dot{m}(h_2 - h_1)$$

The heat input from the boiler is:

$$\dot{Q}_{in} = \dot{m}(h_3 - h_2)$$

Thus the efficiency is:

$$\eta = \frac{(h_3 - h_4) - (h_2 - h_1)}{h_3 - h_2}$$

Note that phase change processes are occurring and that steam does not exhibit ideal gas behavior, so the assumptions of an ideal gas with constant specific heats that were used in gas power cycles cannot be used here with the Rankine cycle. Since the density of liquid water does not change very much, we can approximate the work of the pump by assuming an incompressible liquid for that process, so that:

$$h_2 - h_1 \approx \frac{P_2 - P_1}{\rho}$$

The turbine exhaust pressure is usually about 1/10 of atmospheric. Liquid water in the steam (wet steam) can cause erosion of the turbine blades and reduce the turbine mechanical efficiency. Turbine exhaust temperatures are usually around 40 °C, being limited by the cooling water temperature into the condenser, which is typically at ambient temperatures (around 20 °C) [Deskbook]. Typically is has been practice that up to 10-12% moisture content is allowable at the turbine exit. Higher values of liquid content will result in excessive wear and damage to turbine blades. However modern high-performance turbines tend to be lightweight and more susceptible to damage from liquid drops and any other debris in the steam.

Example 13:
The following properties have been measured at the four states in a Rankine cycle. Compute the efficiency of the cycle.

State	state	h(kJ/kg)	s(kJ/kg-K)	T(°C)	P(bar)
1	Saturate liquid	384.61	1.2137	91.8	0.75
2	Comp. liquid	387.71	1.2138	92.0	30.0
3	Superheat vapor	3116.1	6.7449	350.0	30.0
4	Saturate vapor	2403.0	6.745	91.8	0.75

SOLUTION:

$$\frac{W_{turbine}}{m} = h_3 - h_4 = 3116.1 - 2403.0 = 713.1 \, kJ/kg$$

$$\frac{W_{pump}}{m} = h_2 - h_1 = 387.71 - 384.51 = 3.2\ kJ/kg$$

$$\frac{Q_{in}}{m} = h_3 - h_2 = 3116.1 - 387.71 = 2728.4\ kJ/kg$$

$$\eta \frac{W_{turbine} - W_{pump}}{Q_{in}} = \frac{713.1 - 3.2}{2728.4} = 26.0\%$$

Example 14:
A power plant operates on an ideal Rankine cycle (saturated liquid water compressed isentropically by a pump, vaporized in boiler, expands isentropically through turbine, then goes through condenser). The condenser pressure is 100 kPa and the boiler pressure is 4000 kPa. You may neglect the work of the pump when computing the efficiency. The temperature at the turbine inlet is 600 °C. Calculate the efficiency for the cycle.

SOLUTION: We have the following information, filled out in the table below, where state 2 is compressed liquids and state 3 is superheated vapor. We need to find the enthalpy (h) at each state in order to calculate the efficiency of the cycle. Since we have two properties known at states 1, 3, and 4, and we are using a pure substance (steam = H_2O) as the working fluid, those states are completely defined and we can use any source of steam tables to find the missing information.

State	v(m³/kg)	x	h(kJ/kg)	s(kJ/kgK)	T(°C)	P(bar)
1		0.0			99.6	1.0
2		-				40.0
3		-			600	40.0
4		1.0			99.6	1.0

In this problem I have chosen to use https://webbook.nist.gov/chemistry/fluid/ Selecting **Saturation properties – temperature increments** to get the properties at states 1 and 4, and **isothermal properties** to get state 3, as shown in the next table. To get the properties at state 2, we must use the isentropic compression relation between states 1 and 2, where $s_1 = s_2$.

State	v(m³/kg)	x	h(kJ/kg)	s(kJ/kg-K)	T(°C)	P(bar)
1	0.00104	0.0	417.48	1.3027	99.6	1.0
2		-		1.3027		40.0
3	0.09886	-	3674.87	7.3705	600	40.0
4	1.69426	1.0	2674.94	7.3589	99.6	1.0

Obtaining the properties at state 2 with online steam tables requires some trial and error, and with printed steam tables would require *interpolation*. In this case it quickly becomes apparent that the temperature at state 2 must be very close to the temperature at state 1 for the entropies to match. (Also useful to note: raising the pressure always decreases the entropy, while raising the temperature always increases the entropy, so in raising the pressure from 1 to 40 bars, we know the temperature must be above 99.6 °C.) In this case rounding to the nearest 0.1 °C will be sufficient (most temperature devices cannot measure to any greater accuracy anyway) and we obtain the closest entropy at 99.9 C.

State	v(m³/kg)	x	h(kJ/kg)	s(kJ/kg-K)	T(°C)	P(bar)
1	0.00104	0.0	417.48	1.3027	99.6	1.0
2	0.00104	-	421.68	1.3027	99.9	40.0
3	0.09886	-	3674.87	7.3705	600	40.0
4	1.69426	1.0	2674.94	7.3589	99.6	1.0

An alternative (and probably easier) way to get the enthalpy change across the pump from states 1-2 is to use the definition of enthalpy: h = u + Pv.

$$\Delta h = (u_2 + P_2 v_2) - (u_1 + P_1 v_1)$$

As we can see in the above table, to the number of significant digits reported, the specific volume is constant, so $v_1 = v_2$. Also the internal energy depends on temperature much more strongly than it does on pressure, and there is very little temperature change across the pump, so we can approximate that $u_2 \approx u_1$. Thus the change in enthalpy is:

$$\Delta h \approx (P_2 - P_1)v = (40 - 1 \text{ bar})\left(0.00104 \frac{m^3}{kg}\right) = 4.16 \frac{kJ}{kg}$$

The actual tabular data gives Δh = 4.2 kJ/kg, and as we shall soon see in the context of this problem that is sufficiently accurate, as 0.04 kJ/kg will not make a significant difference in the overall energy balance. To calculate the efficiency of this Ranking cycle we need to calculate the energy input per mass of steam and the next work generated per mass of steam. The net work is the work of the turbine minus the work of the pump.

$$\frac{W_{turbine}}{m} = h_3 - h_4 = 3674.87 - 2674.94 = 999.9 \text{ kJ/kg}$$

$$\frac{W_{pump}}{m} = h_2 - h_1 = 421.68 - 417.48 = 4.2 \text{ kJ/kg}$$

$$\frac{Q_{in}}{m} = h_3 - h_2 = 3674.87 - 421.68 = 3253.2 \; kJ/kg$$

$$\eta \frac{W_{turbine} - W_{pump}}{Q_{in}} = \frac{999.9 - 4.2}{3253.2} = 30.6\%$$

How does the efficiency of the basic Rankine cycle vary with boiler operating pressure? Assumed fixed operating points of saturated liquid at state 1 and saturated vapor at state 4, with a fixed condenser operating pressure, then as the boiler pressure increases, the boiler temperature also increases, producing a larger pressure and temperature drop across the turbine, which will result in an increase in efficiency of the cycle. (This would be a good homework assignment to plot the efficiency of the cycle as a function of boiler pressure).

Carnot Cycle

The expensive thermo books would have covered the Carnot cycle well before now, but I have omitted it because I do not feel it represents any real device.

When we look at the previous two examples, you might think the efficiencies are quite low. How could we improve the efficiency? We could add a second low pressure turbine after the main turbine to extra more energy out of the steam before it goes to the condenser. We could even add a third turbine, or use regenerators, reheat, or open feedwater heaters to further increase the efficiency, but there is a limit on the maximum efficiency. The second law of thermodynamics tells us that there must always be some waste heat rejected in any process, so that we can never extract all the energy from the working fluid, and the efficiency must be less than 100%. In the case of a Rankine power cycle, eventually the steam must go through the condenser and give up its remaining available energy to the environment.

The **Carnot efficiency** is the maximum efficiency obtainable for an engine (work generating device) operating between two sources or sinks of heat – one a high temperature source at T_H from which energy is extracted, and the second a low temperature sink at T_C to which heat is exhausted. The maximum possible efficiency is obtained when the heat transfers from the system to the hot and cold reservoirs are proportional to the temperature ratio of those reservoirs:

$$\frac{Q_C}{Q_H} = \frac{T_C}{T_H}$$

For simple cycle in which heat is added to the system from a hot source, Q_H, work (W) is done, and heat is exhausted to the cold environment (Q_C), the first law energy balance is:
$W = Q_H - Q_C$. The efficiency of such a system is:

$$\eta = \frac{W}{Q_H} = \frac{Q_H - Q_C}{Q_H}$$

Combining the previous two equations, we obtain the efficiency of the theoretical Carnot cycle:

$$\eta_{carnot} = 1 - \frac{T_C}{T_H}$$

where both T_C and T_H must be absolute units of Kelvin (or Rankine, but seriously, who uses temperatures in Rankine?)

To create a Carnot engine, we can construct a theoretical Carnot Cycle, consisting of four thermodynamic processes, shown in Figure 23:
1. Isothermal Compression
2. Isentropic Compression
3. Isothermal Expansion
4. Isentropic Expansion

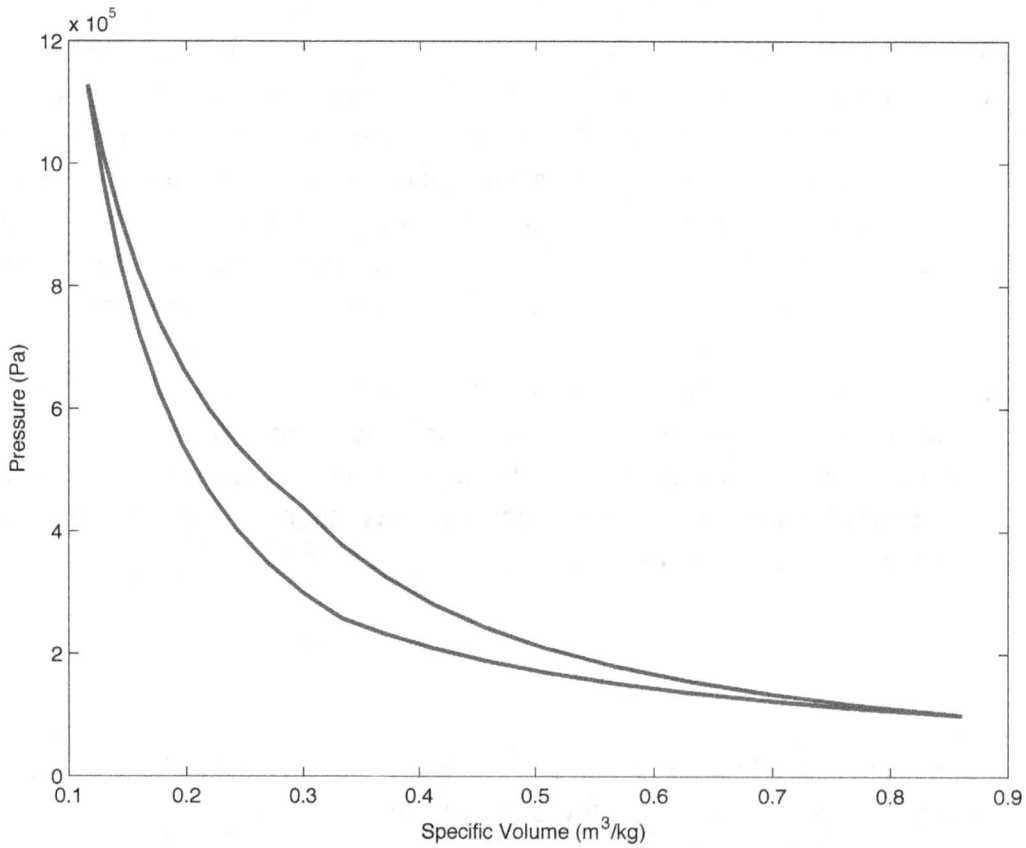

Figure 23: P-V diagram of Carnot cycle for a heat engine.

Figure 23 hints at why the theoretical Carnot cycle does not make a useful model for practical thermal engines. The net work produced in a cycle can be calculated graphically from a P-V diagram, using the integral relation:

$$W = \int P \, dV$$

In other words, the area bound inside the blue lines in Figure 23 is the net work of the cycle. The area is quite small, due to the narrow shape. Compare this to Figure 26 for the Brayton cycle which models gas turbines, and has a large area under the curve. So the Carnot cycle represents the cycle that has the highest possible efficiency, but at the cost of only performing a small amount of work (and doing it very slowly, too).

In any case, we can still use the Carnot efficiency to tell us the maximum attainable efficiency in a given situation. For the Example #13 problem, T_H = 350 °C = 623 K, and T_C = 92 °C = 365 K. So the Carnot efficiency is:

$$\eta_{carnot} = 1 - \frac{T_C}{T_H} = 1 - \frac{365 \, K}{623 \, K} = 41.4\%$$

I do not find the Carnot cycle to be of much use, since no device in real-life comes anywhere close to Carnot operating conditions. The Carnot cycle does provide an upper bound on the maximum efficiency you could obtain for and engine or power plant, but since no device comes anywhere near that limit, I do not think it very useful. However, it does provide insight into trends in efficiency as you change temperature.

Endoreversible Carnot Cycle

There is another theoretical cycle, called the **endoreversible Carnot cycle**, which gives a more realistic approximation for the obtainable efficiency for real processes. The endoreversible Carnot Cycle differs from the Carnot cycle in that it considers heat transfer across a finite temperature difference between the hot and cold reservoirs to the working fluid of the cycle, along with a corresponding *thermal resistance* to heat transfer from the heat reservoirs to the working fluid. These heat transfer losses result in a lower calculated thermodynamic efficiency than the Carnot cycle, but more realistically model real processes. For the same conditions as the previous problem, the theoretical efficiency is calculated as:

$$\eta_{carnot,endo} = 1 - \sqrt{\frac{T_C}{T_H}} = 1 - \sqrt{\frac{365 \, K}{623 \, K}} = 23.5\%$$

compared to 26.0% calculated from the Rankine cycle.

Example 15:

The Geysers geothermal plant in California provides steam at about 180 °C and 0.8 MPa, with a flow of two million pounds of steam per hour about 250 kg/s) to operate a 110 MW generating unit. The turbine outlet is saturated vapor at 33 °C. Calculate the potential efficiency of the plant assuming a Carnot cycle, and a basic Rankine cycle.

SOLUTION: For this geothermal power problem, T_H = 180 °C = 453 K, and T_C = 33 °C = 306 K. So the Carnot cycle the efficiency is:

$$\eta_{carnot} = 1 - \frac{T_C}{T_H} = 1 - \frac{306\ K}{453\ K} = 32.5\%$$

The **endoreversible Carnot cycle** gives a more realistic upper bound:

$$\eta_{carnot,endo} = 1 - \sqrt{\frac{T_C}{T_H}} = 1 - \sqrt{\frac{306\ K}{453\ K}} = 17.8\%$$

For an equivalent Rankine Cycle, the heat has already been provided from the geothermal reservoir, so the conditions at state 3 are P_3 = 0.8 MPa = 8 bar, and T_3 = 180 °C. We have to assume something about the outlet condition of the turbine (state 4) since no other information was provided. We also assume state 1 is saturated liquid at the same pressure as the turbine outlet. Note this pressure is below atmospheric pressure. The lower the pressure, the lower the temperature (for a saturated fluid) and the higher efficiency that can be obtained. The limitations on how the pressure can go are:
- That the temperature must still be higher than the surrounding atmospheric temperature (for heat rejection to occur in the condenser)
- That the condenser is now at negative pressure compared to the surrounding atmosphere, (all piping must be well sealed to prevent any leaks of outside air from coming in and contaminating the working fluid).
- Also there is no well-defined state 2 in this cycle, since pressurized fluid is extracted from the reservoir.

The thermodynamic data we have thus far for modeling the process as a basic Rankine cycle is shown in the following table:

State	Pressure (bar)	Temperature (°C)	Quality	Enthalpy (kJ/kg)	Entropy (kJ/kg-K)
1	0.05	33	0.0	138.3	0.4779
2	8				
3	8	180	superheat	2792.4	6.7154
4	0.05	33	1.0	2567.0	8.3913

The work per unit mass obtained from the turbine is: $\Delta h = h_3 - h_4 = 225.4$ kJ/kg

We can approximate the heat input to the system as: $\Delta h = h_3 - h_1 = 2654.1$ kJ/kg

Thus the thermodynamic efficiency is:

$$\eta = \frac{W}{Q} = \frac{225.4}{2654.1} = 8.5\%$$

Geothermal power plants typically suffer from low thermodynamic efficiencies due to the relatively low temperature of the geothermal fluid used. The efficiency can be improved by using multi-stage turbines. Typically efficiencies of geothermal power plants observed in practice are around 10-17%.

In comparison to a coal-fired thermal power plant, a typical nuclear power plant steam is at 290 °C and 7.5 MPa. [Deskbook], This is a bit cooler due to material limitations of piping going through the reactor that must be able to withstand the radiation from the fission core. For energy sources at relatively low temperature, such as wet geothermal system or solar thermal power, it is necessary to use a fluid with a lower critical point than water. Organic Rankine Cycles (ORC) are often used for this purpose, with the working fluid often being a lightweight hydrocarbon such as pentane (hence the name organic) or some refrigerants are also suitable. See Table 17 in section on refrigerators for the properties of common refrigerants compared to water.

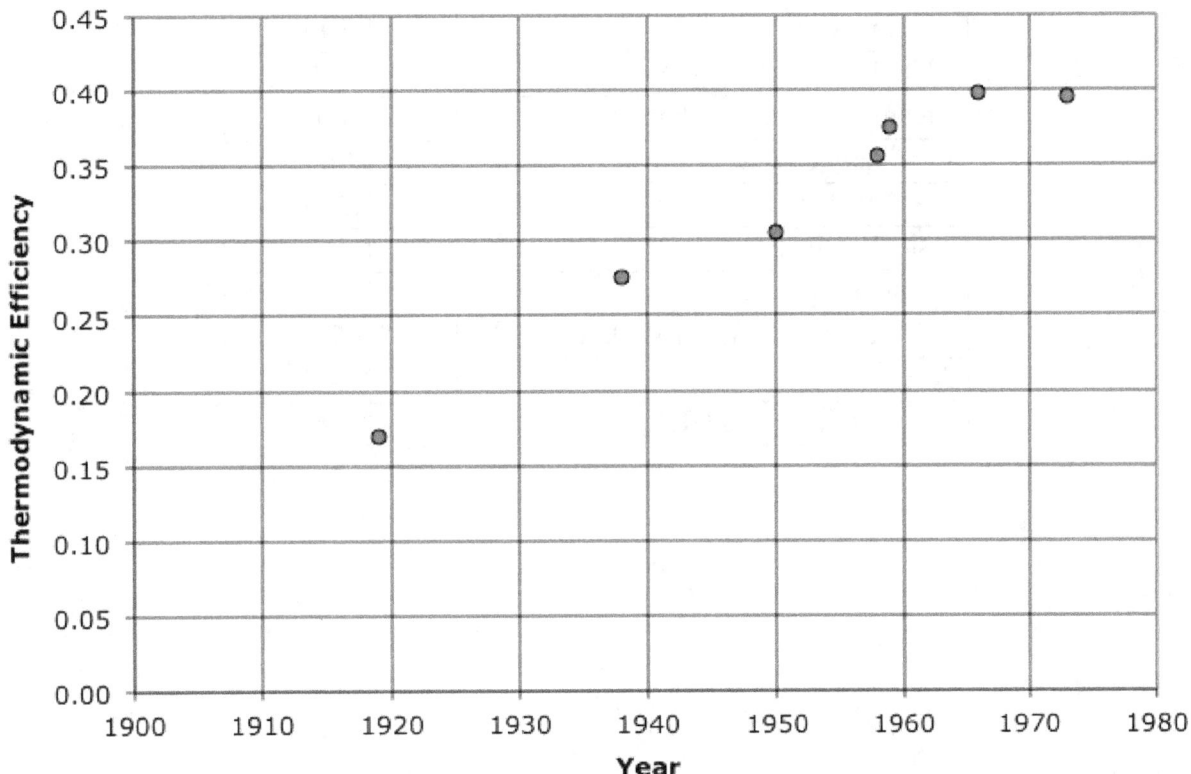

Figure 24: Historical Efficiency of UK fossil fuel power plant stations. Data from: Analysis of Engineering Cycle. R. Haywood. Pergamon Press. 1967

Table 14: Summary of auxiliary loads on a 400 kWe pulverized coal power plant operating at 100% load, steam at P = 2400 psi (16.5 Pa) and T = 1000 °F (540 °C). Balance of plant includes control systems, lighting, HVAC, etc. Data from [DOE99]

Load Source	Power (kWe)
Gross Power	**422,224**
Losses	
Coal Handling	200
Limestone Handling & Reagent Prep	850
Pulverizers	1,730
Condensate Pumps	780
Main Feed Pump	8,660
Miscellaneous Balance of Plant	2,000
Primary Air Fans	1,000
Forced Draft Fans	1,000
Induced Draft Fans	4,302
Seal Air Blowers	50
Precipitators	1,100
FGD Pumps and Agitators	3,200
Steam Turbine Auxiliaries	700
Circulating Water Pumps	3,360
Cooling Tower Fans	1,900
Ash Handling	1,550
Transformer Loss	1,020
TOTAL AUXILIARIES, kWe	24,742
Net Power, kWe	**397,482**
Net Efficiency, % HHV	37.6

Gas Turbine

If we combine a compressor, a combustor, and a turbine in series, we create a gas turbine engine or power plant.

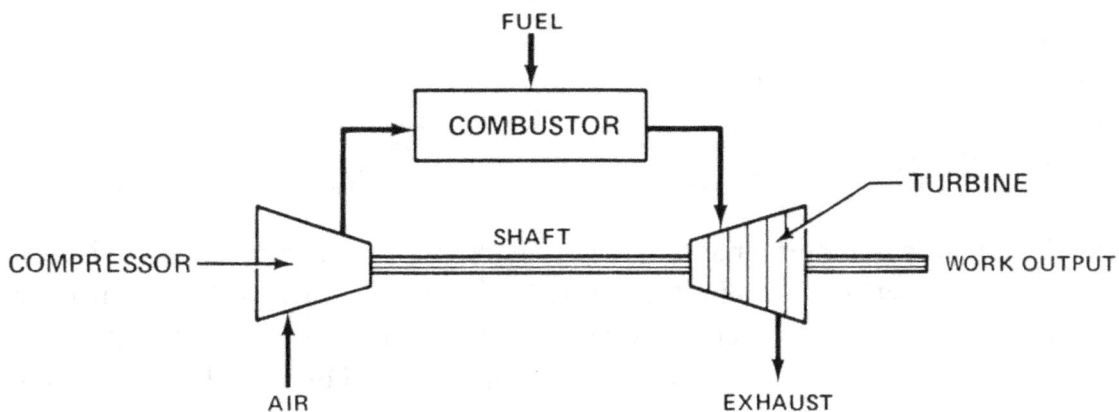

Figure 25: Components of a basic Brayton Cycle gas turbine. [Deskbook]

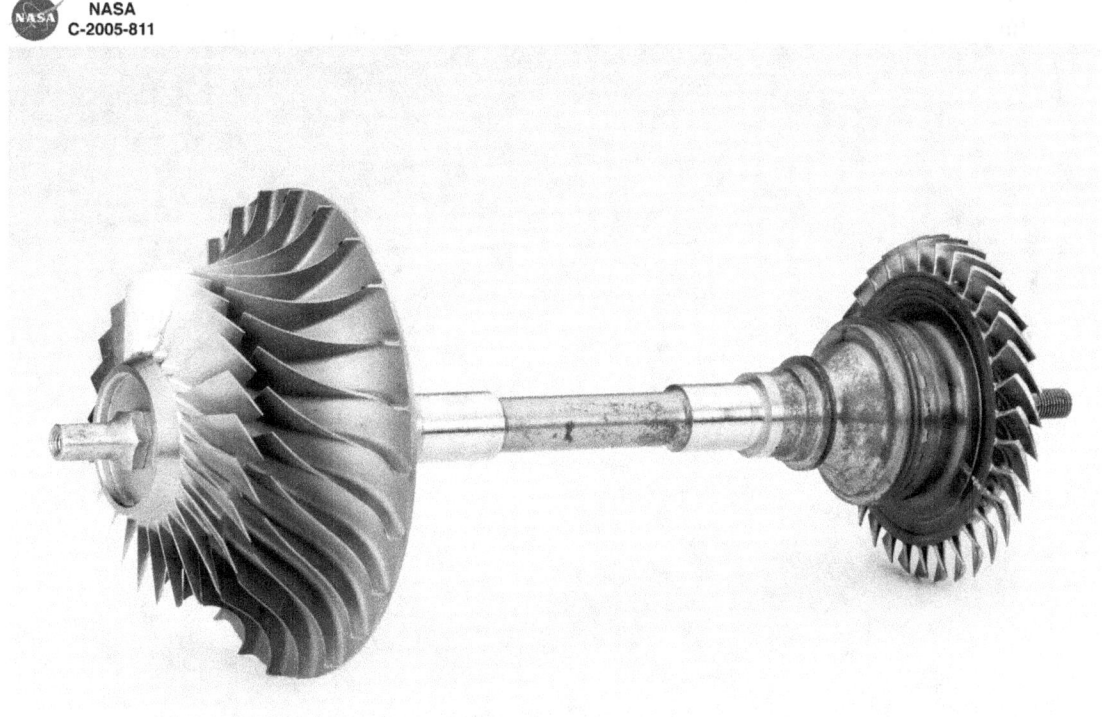

Figure 26: Compressor and Turbine. Source NASA Glenn Image Archives.

The Brayton Cycle has come to be used to model open flow gas turbine engines, with continuous combustion.

Table 15: Steps in Brayton Cycle:

Process	Thermodynamic relations	Pressure relations
Isentropic Compression	$T_2 = T_1*(P_2/P_1)^{(k-1)/k}$	$P_2 = P_1 *$ Compressor P ratio
Const. Pressure Heat Addition	$Q_{IN} = m\, c_P\, (T_3 - T_2)$	$P_3 = P_2$
Isentropic Expansion	$T_4 = T_3*(P_4/P_3)^{(k-1)/k}$	$P_4 = P_3 /$ Compressor P ratio
Const. Pressure Heat Rejection	$Q_{OUT} = m\, c_P\, (T_4 - T_1)$	$P_4 = P_1$

It can be shown the efficiency of the simple Brayton cycle increases as the pressure ratio increases. An important parameter in analyzing gas turbines is the back work ratio, which is the percentage of work generated by the turbine that goes to power the compressor. Gas turbines can be used either for aircraft propulsion or stationary power generation. In propulsion applications all of the turbine power must go to the compressor or other thrust generating devices, such as fans or propellers. More complex Brayton cycles for power plants can be constructed when the use of regenerators or multiple turbines is to be modeled. Coal-burning and nuclear power plants are normally modeled with a Rankine cycle. However, this involves phase change of the working fluid and therefore is not included in this book.

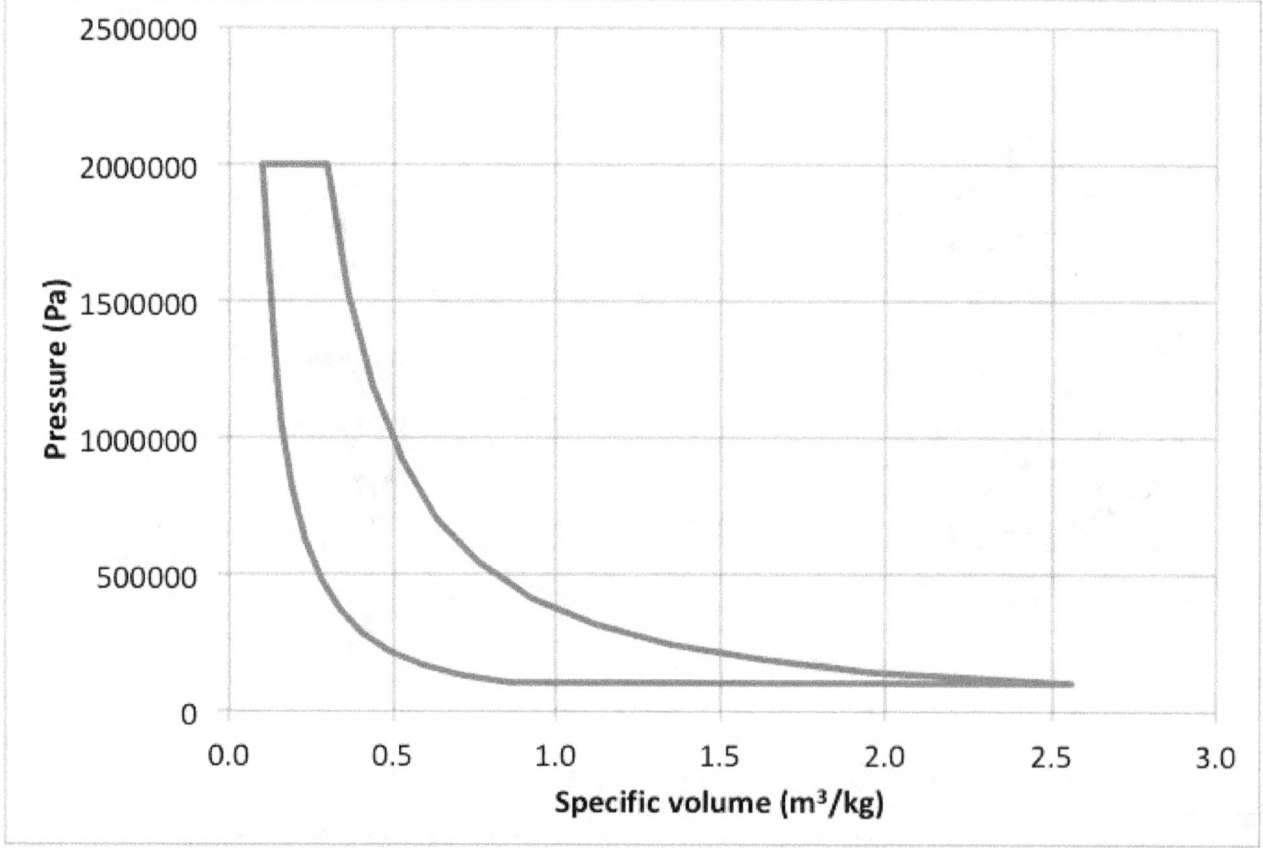

Figure 27: P-V diagram of Brayton cycle with a compressor pressure ratio of 20 and heat addition of 1400 kJ/kg.

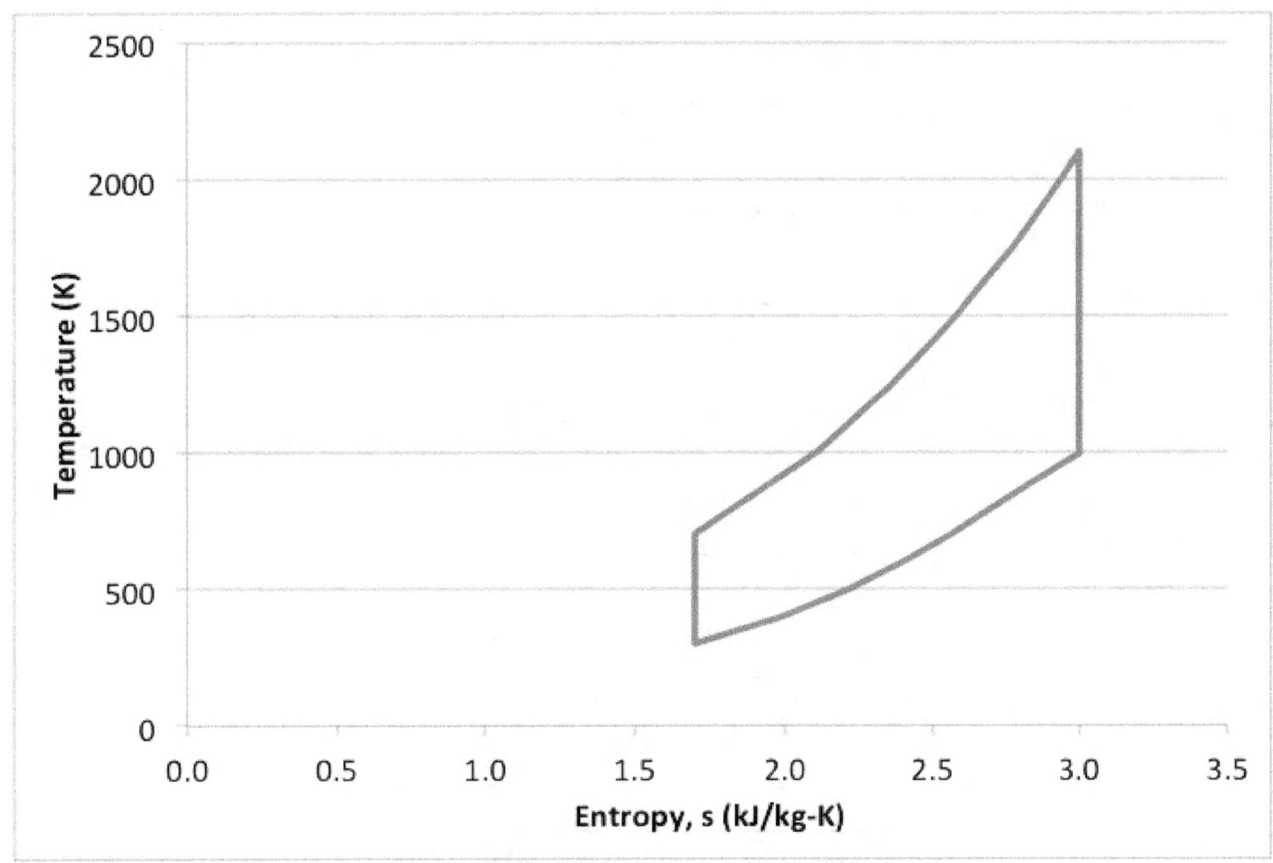

Figure 28: T-s diagram of Brayton cycle

Example 16:
Compute the efficiency and the states of the simple Brayton cycle with a pressure ratio of 20 and heat addition of 1400 kJ/kg.

SOLUTION: Since the compression process is assumed to be isentropic, we can use the isentropic relations for an ideal gas to calculate the temperature at the compressor exit:

$$T_2 = T_1 \left(\frac{P_2}{P_1}\right)^{(k-1)/k} = 300\ K\ (20)^{(1.4-1)/1.4} = 706\ K$$

Assuming a constant pressure combustor, and constant specific heats, the temperature at the outlet of the combustor can be calculated from:

$$T_3 = T_2 + \frac{Q/m}{c_P} = 706\ K + \frac{1400\ kJ/kg}{1.005\ kJ/kg \cdot K} = 2099\ K$$

The turbine expands the exhaust gases out to atmospheric pressure, so the final temperature can be calculated as:

$$T_4 = T_3 \left(\frac{P_4}{P_3}\right)^{(k-1)/k} = 2099\,K \left(\frac{1}{20}\right)^{(1.4-1)/1.4} = 892\,K$$

The heat rejected to the environment can be calculated:

$$\frac{Q_{OUT}}{m} = c_P(T_3 - T_3) = 1.005\,\frac{kJ}{kg \cdot K}(892\,K - 300\,K) = 595\,kJ/kg$$

The efficiency can be calculated from an energy balance as:

$$\eta = \frac{1400 kJ/kg - 595\,kJ/kg}{1400\,kJ/kg} = 0.575 = 5.5\%$$

Table 16: Results for Brayton Cycle Example 16.

State	Pressure (bar)	Temperature (K)
1	1.0	300
2	20.0	706
3	20.0	2099
4	1.0	892

Refrigerators

It is fairly easy to heat things up. As an example one could use electrical resistance in wires for heating air, or burn a fuel to release heat. Making things colder is more difficult. To take away heat from an object, we must place in contact with another substance that is even colder than the desired temperature, in order to induce heat transfer in the desired direction. One way to do this is to use the latent heat of vaporization, and mechanical induce a phase change to cause the desired cooling. A good **refrigerant** is a fluid that will vaporize at the desired pressure and have a large value of the heat of vaporization, h_{fg}, at that temperature, to reduce the mass of the refrigerant needed to get the desired amount of cooling.

For proper design of a refrigeration cycle, there are a few requirements:
- In order for heat transfer to occur in the condenser, the saturation temperature of the working fluid (the refrigerant) at the condenser pressure must be greater than the outside hot air temperature (sometimes referred to as the temperature of the high-temperature reservoir)
- In order for heat transfer to occur in the evaporator, the saturation temperature of the working fluid (the refrigerant) at the evaporator pressure must be less than the inside cold air temperature (sometimes referred to as the temperature of the low-temperature reservoir)
- The working fluid (refrigerant) should be selected so that (1) the saturation pressure in the condenser is not excessively high (which would require components with thick walls to withstand the mechanical stresses), (2) the saturation pressure in the evaporator is not too low – (ideally above atmospheric pressure, so the system maintains positive pressure and there is no possibility of air leaking into the refrigerant loop)
- Ideally the working fluid is not corrosive or toxic
- The heat of vaporization (latent heat, h_{fg}) should be high at the temperature of the evaporator.
- Usually we want to select a fluid-system combination that has fully vaporized before entering the compressor, to avoid liquid drops causing damage to the compressor blades.

Table 17 shows a selection of commonly used refrigerants. Note that the critical pressure and temperature of water are much higher than the fluids used as refrigerants. R12 is also known as Freon-12 or CFC-12, but its use has been banned by the Montreal Protocol to prevent the destruction of stratospheric ozone that occurs when the chlorine from the chlorofluorocarbon molecules is released in the atmosphere.

Table 17: Relevant properties of some fluids commonly used as refrigerants, compared to water.

Name	T_{fus} (°C)	T_{bn} (°C)	T_{crit} (°C)	P_{sat} at 0 °C (bar)	P_{sat} at 50 °C (bar)	P_{crit} (bar)	h_{fg} @ P_{atm} (kJ/kg)
Water	0	100	375	0.0	0.1	220	2260
Ammonia	-78	-33	132	4.3	20.3	113	1357
SO$_2$	-72	-10	157	2.0	9.0	78.8	381
CO$_2$ (R744)	-78	-78	31	35.0	60.0	73.8	230
R22	-160	-41	96	5.0	19.4	49.8	233
Propane	-187	-42	97	4.7	17.1	42.5	356
R12	-158	-30	112	3.1	12.2	41.3	165
R134a	-103	-27	101	2.9	13.0	40.6	218
Butane	-137	0	152	1.0	5.0	38.0	385
R245fa	-73	15	154	0.5	3.5	36.4	184
Pentane	-130	36	197	0.2	1.6	33.6	370

Figure 29 shows the simple 4-step refrigeration cycle, with its 4 components.

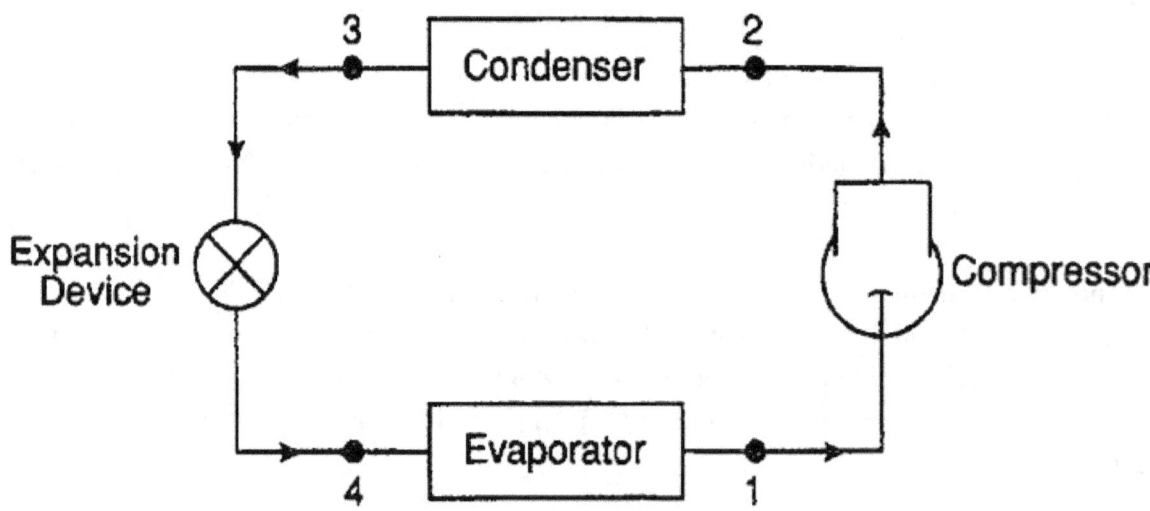

Figure 29: Basic 4-step refrigeration cycle. From [NIST95]

Figure 30: Cooler for large compressors. From NASA Glenn Image Archives.

Example 17:

An ideal **Refrigeration** cycle with R134a as the working fluid consists of isentropic compression from saturated vapor at -10 °C (state 1) to a superheated state at 34 °C (state 2), constant-pressure heat rejection in the condenser to the condenser exit, at saturated liquid (state 3), constant-enthalpy pressure loss across a throttle valve to saturated liquid-vapor mixture (state 4), and constant-pressure heat absorption in the evaporator to a saturated vapor (back to state 1). Calculate the coefficient of performance of the cycle and fill out the data in the table.

SOLUTION:

Using the NIST Chemistry Webbook (https://webbook.nist.gov/chemistry/fluid/) select R134a as the species of interest, select the appropriate units, and saturation properties – temperature increments, to get the data at states 1, 3, and 4. So to start with, these are the known states of the working fluid and thermodynamic relations, arranged in a table:

Table 18: Initial thermodynamic property data for basic refrigeration cycle.

State	Pressure (bar)	Temperature (°C)	Quality	Enthalpy (kJ/kg)	Entropy (kJ/kg-K)
1		-10	1.0		
2		34	vapor		$s_2 = s_1$
3	$P_3 = P_2$		0.0		
4	$P_4 = P_1$	-10		$h_4 = h_3$	

We already have 2 defined properties (T, x) at state 1, which defines the state and allows us to look up the remaining properties (P_{sat}, h_g, s_g).

State	Pressure (bar)	Temperature (°C)	Quality	Enthalpy (kJ/kg)	Entropy (kJ/kg-K)
1	2.006	-10	1.0	392.66	1.7334
2		34	vapor		1.7334
3	$P_3 = P_2$		0.0		
4	2.006	-10		$h_4 = h_3$	

Now to find state 2, we know two properties (T,s) so the state is defined, but it is bit more difficult to find the remaining properties. We know it will be a superheated vapor state, but we need to find the pressure in order to find the enthalpy from data tables and most databases. Using the NIST Chemistry Webbook again, this time use the isothermal properties. Now we will have to iterate (guess) to find the pressure. We know pressure increases with temperature, so $P_2 > 2$ bar. In this case eventually you will find the entropy matches with P = 7.4 bar

State	Pressure (bar)	Temperature (°C)	Quality	Enthalpy (kJ/kg)	Entropy (kJ/kg-K)
1	2.006	-10	1.0	392.66	1.7334
2	7.4	34	-	419.75	1.7334
3	7.4		0.0		
4	2.006	-10	??		

Now state 3 is defined (P,x known), so using the saturation properties we can find T_3 and h_3.

State	Pressure (bar)	Temperature (°C)	Quality	Enthalpy (kJ/kg)	Entropy (kJ/kg-K)
1	2.006	-10	1.0	392.66	1.7334
2	7.4	34	-	419.75	1.7334
3	7.4	28.6	0.0	239.72	1.1370
4	2.006	-10	??	239.72	

The work input required is: W/m = $h_2 - h_1$ = 419.75 - 392.66 = 27.09 kJ/kg
The cooling load provided is: Q/m = $h_1 - h_4$ = 392.66 - 239.72 = 152.94 kJ/kg

The coefficient of performance of the refrigeration cycle is: β = Q/W = 152.94/27.09 = 5.6.

Note: to find the quality at state 4, we need to find the saturation properties at -10 °C. From the NIST chemistry Webbook, h_f = 186.70 and h_g = 392.66 kJ/kg. Therefore the quality is:

$$x = \frac{239.72 - 186.70}{392.66 - 186.70} = 0.26$$

The coefficient of performance of a Carnot refrigerator operating between the maximum and minimum temperatures of the cycle is:

$$\beta_{Carnot} = \frac{T_C}{T_H - T_C} = \frac{263\ K}{307\ K - 263\ K} = 6.0$$

The efficiency of the ideal vapor-compression refrigeration cycle is very close to the efficiency of Carnot refrigerator for this example. What are the reasons why a real refrigerator would be less efficiency than this ideal vapor-compression refrigeration cycle?
- Friction losses due to fluid flow in the pipes, results in pressure not being constant in either condenser or evaporator
- Non-isentropic compressor
- Heat transfer losses in the piping between components of the cycle

This problem would be much easier with EES or another software package that can solve non-linear equations, or a database that allows you to specify entropy as one of the state variables to find the properties. Other resources you could have used to find the properties of refrigerant R134a are:
https://www.ohio.edu/mechanical/thermo/property_tables/R134a/index.html
http://www.peacesoftware.de/einigewerte/r134a_e.html

Note different databases may have different values for *h* & *s* due to the choice of reference data point for 0. Since we are usually only interested in changes of enthalpy and/or entropy this is fine, but you often can't combine data from different sources.

Example 18

A heat pump using R32 (T_c = 78.4 °C, T_{bn} = -51.6 °C) as the working fluid requires 0.66 kW of electricity to run (mainly to power the compressor). In air-conditioning mode it can provide 2.5 kW of cooling, and in heat pump mode it can provide 3.2 kW of heating. Find the coefficients of performance as both an air conditioner and a heat pump, and the cost savings when used as a heat compared to an electric resistance heater.

SOLUTION: The coefficient of performance of a refrigerator is:
$$\beta = \frac{Q_C}{W} = \frac{2.5\ kW}{0.66\ kW} = 3.8$$
The coefficient of performance of a heat pump is:
$$\gamma = \frac{Q_H}{W} = \frac{3.2\ kW}{0.66\ kW} = 4.8$$
Also note $Q_H = Q_C + W$, so $\gamma = \beta + 1$.

An electric heater providing the same heating load would require 3.2 kW of electricity. So if electricity costs $0.24/kW-hr, and the electric heating was used for 1000 hours over the winter, the cost to run it would be:

$$(3.2\ kW)(1000\ hours)(\$0.24/kW\text{-}hr) = \$768$$

The cost to run the heat pump would be:

$$(0.66\ kW)(1000\ hours)(\$0.24/kW\text{-}hr) = \$158$$

A savings of $610 per 1000 hours of operation.

Note that in countries using 240 V electricity voltage, the most power a single plug-in wall heater could produce is: (240 V)(10 A) = 2400 W = 2.4 kW. So either multiple heaters would be needed to provide the same heating capacity as the heat pump, or the electric heater would need to be cycled on more frequently during each day.

Tank filling and emptying

Many thermodynamic problems are concerned with the steady-state operation of devices such as power plants or engines, but there are also some important transient problems, such as tank filling and tank emptying problems. In an **open system**, mass can flow freely in and out of the stems or valves. The most general form of conservation of mass is that the mass in the control volume changes depending in the inflows and outflows of mass to the system:

$$\frac{dm_{CV}}{dt} = \sum \dot{m}_{in} - \sum \dot{m}_{out}$$

For steady-state, steady flow problems, which often applies to turbines, pumps, compressors, nozzles, and diffusers, among other devices, we assume steady-state so that dm/dt = 0, and the total mass flowing into the control volume equals the total mass exiting the system. For most of the problems in this book the assumption of steady-state has been used. In this section we will look at transient problems where the mass in the control volume does change. For open systems, the most general form of conservation of energy is:

$$\frac{dE_{CV}}{dt} = \sum \pm \dot{Q} + \sum \pm \dot{W} + \sum_{in} \dot{m}\left(h + \frac{V^2}{2} + gz\right) - \sum_{out} \dot{m}\left(h + \frac{V^2}{2} + gz\right)$$

The energy in the control volume is the internal energy:

$$E_{CV} = m_{CV} e$$

where e is the specific energy of the fluid in the control volume. This energy is the sum of the internal (thermal) energy and the kinetic and potential energy per unit mass.

$$e = u + \frac{V^2}{2} + gz$$

Some common examples of unsteady problems where the steady-state assumption will not apply are emptying and filling of vessels, in addition to the transient startup of devices. For the problems of tank filling and emptying, no mechanical work is being done and the changes in potential energy can be neglected. With the negligible kinetic and potential energy, the specific energy is simply equal to the internal energy, so that:

$$e = u$$

Note that one exception where this assumption would not be valid is high a high pressure vessel were suddenly pierced so that the fluid exited through an uncontrolled leak, the kinetic energy of the escaping gas would be quite important, particularly to anyone standing nearby. With the assumptions of no mechanical work being done and negligible kinetic and potential energy, the first law of thermodynamics simplifies to:

$$\frac{d(mu)_{CV}}{dt} = \sum \pm \dot{Q} + \sum_{in} \dot{m}h - \sum_{out} \dot{m}h$$

Tank filling

For a tank filling problem we assume no mass is leaving the tank and that there is a single inflow of mass into the tank, so that the conservation of energy equation simplifies to:

$$\frac{d(mu)_{CV}}{dt} = \dot{Q} + \dot{m}_{in} h_{in}$$

Typically the inflow is coming from a large supply so that we can treat h_{in} as a constant. Then we can integrate this equation (a differential equation) to obtain an algebraic equation:

$$m_2 u_2 - m_1 u_1 = \dot{Q} \Delta t + \dot{m}_{in} h_{in} \Delta t$$

where state 1 represents in the initial state of the system and state 2 the final state, and the filling of the vessel takes place over a time Δt. The change in mass of the tank is calculated by:

$$m_2 - m_1 = \dot{m}_{in} \Delta t$$

If we further assume that the tank was initially empty (vacuum tank), so that $m_1 = 0$, then conservation of energy simplifies to:

$$u_2 = \frac{Q}{m_2} + h_{in}$$

Furthermore, if the tank is insulated so that $Q = 0$, then the final internal energy in the tank will be equal to the enthalpy of the influx stream:

$$u_2 = h_{in}$$

Since $h = u + Pv = u + P/\rho$, this requires that the final temperature in the tank be greater than the temperature of the inflowing fluid stream. The final pressure in the tank, P_2, will be less than or equal to the supply pressure, P_{in}.

Example 19:
An initially empty tank is being filled with steam from a constant supply at P = 0.50 MPa and T = 500 °C. The tank is allowed to fill until the final pressure in the tank is P = 0.50 MPa. Calculate the final temperature of the steam in the tank, assuming the tank is insulated so there are no heat losses.

SOLUTION: Equation # applies, so that $u_2 = h_{in}$. Using a suitable electronic resource for water properties, such as:
http://www.steamtablesonline.com/steam97web.aspx
The enthalpy of the incoming steam at P = 0.50 MPa, T = 500 °C, is h = 3484.5 kJ/kg. (If you are using tables in your textbook, you will need to find in the same table the temperature at which u = 3484.5 kJ/kg and P = 0.50 MPa. Linear interpolation must be used, yielding a final temperature of **T_2 = 703 °C**). Note that this temperature is higher than the supply temperature, since the flow work (Pv) of the incoming steam gets converted into thermal energy when it expands into the tank.

Tank emptying

For problems of tank emptying, we can assume no new mass will be entering the tank. If we further assume only a single outflow, then the Conservation of Energy equation simplifies to:

$$\frac{d(mu)_{CV}}{dt} = \dot{Q} - \dot{m}_{out} h_{out}$$

In the most general case the h_{out} changes as the u_{CV} changes, and P drops as mass leaves the tank (such as for an ideal gas), and we have to solve a transient differential equation. If however the h_{out} is constant, then the integration becomes simpler and we have:

$$m_2 u_2 - m_1 u_1 = \dot{Q}\Delta t - \dot{m}_{out} h_{out} \Delta t$$

This might apply if we have a saturated two-phase mixture in the tank, and either saturated liquid at constant temperature is drained from the bottom of the tank, or saturated vapor at constant temperature is removed from the top of the tank. For such a saturated mixture the *quality*, X, of the mixture in the tank will change with time (usually increasing as the mixture becomes more vapor and less liquid as mass is removed). Another possibility for a tank filled only with a gas is that it could be maintained in an *isothermal* environment, where Q is added to keep the temperature, T, in the tank a constant. For such cases, recognizing that $m_2 - m_1 = \dot{m}_{out}\Delta t$, this equation simplifies even further.

$$m_2 u_2 - m_1 u_1 = \dot{Q}\Delta t - (m_2 - m_1) h_{out}$$

For the special case of an *adiabatic* process where the previous assumptions apply:

$$m_2 u_2 - m_1 u_1 = -\dot{m}_{out}(m_2 - m_1)h_{out}$$

Example 20:
A rigid tank of volume 10 L is initially filled with a saturated steam-water mixture at 205 kPa (absolute). Initially the vessel contains 50% steam and 50% water (by volume). Heat is added to the mixture at a rate of 1 kW. A pressure release valve on the top vents steam to maintain a constant pressure of 205 kPa so that the tank does not over-pressurize and explode. How long will it take until all the liquid water in the tank has vaporized?

SOLUTION: A suitable electronic resource for steam properties (such as http://www.steamtablesonline.com/steam97web.aspx) can be used to find the property data.
- The boiling point at 200 kPa is T_{sat} = 121 °C.
- The densities of saturated liquid and vapor are 942.3 kg/m³ and 1.156 kg/m³, respectively.
- The enthalpy of the saturated vapor is h_g = 2707.4 kJ/kg, and the heat of vaporization is h_{fg} = 2199.3 kJ/kg.
- The internal energy of saturated liquid and vapor are u_f = 507.8 kJ/kg and u_g = 2530.0 kJ/kg, respectively.

Since there are 5 L = 0.005 m³ for both liquid and vapor states, then the initial mass of liquid is:

$$m_f = (0.005 \text{ m}^3)*(942.3 \text{ kg/m}^3) = 4.711 \text{ kg}$$

and the initial mass of vapor is:

$$m_g = (0.005 \text{ m}^3)*(1.156 \text{ kg/m}^3) = 0.006 \text{ kg}$$

The total mass in the container at the initial state is: 4.711 kg + 0.006 kg = 4.717 kg.

The mixture quality is:
$$x = m_{vapor}/m_{total} = (0.006 \text{ kg})/(4.717 \text{ kg}) = 0.0013$$

The average internal energy for the initial mixture is:

$$u = x u_g + (1-x) u_f = 0.0013(2530 \text{ kJ/kg}) + 0.9987(507.8 \text{ kJ/kg}) = 510.4 \text{ kJ/kg}$$

At the final state, the container is completely full of saturated vapor, having a mass of

$$m = \rho V = (0.010 \text{ m}^3)*(1.156 \text{ kg/m}^3) = 0.01156 \text{ kg}$$

Thus the mass leaving the system is: $m_2 - m_1 = 4.717 \text{ kg} - 0.012 \text{ kg} = 4.705 \text{ kg}$.

Now we have all the data we need to use the equation:

$$m_2 u_2 - m_1 u_1 = \dot{Q}\Delta t - (m_2 - m_1)h_{out}$$

The only unknown in the equation is the time for the process, Δt.

$$\Delta t = \frac{m_2 u_2 - m_1 u_1 + (m_2 - m_1)h_{out}}{\dot{Q}}$$

$$\Delta t = \frac{(0.012 \text{ kg})\left(2530\frac{kJ}{kg}\right) - (4.717 \text{ kg})\left(510.4\frac{kJ}{kg}\right) + (4.705 \text{ kg})\left(2707.4\frac{kJ}{kg}\right)}{1 \text{ kW}} = 10,361 \text{ s}$$

It will take almost 3 hours for all the liquid water in the tank to vaporize.

Ideal Gases with Constant Specific Heats

Many applications use air at pressures low enough that we can treat it as an ideal gas. Furthermore if the temperature changes are moderate, we can also take the specific heats as constant, to good approximation, so that:

$$u_2 - u_1 = c_V (T_2 - T_1)$$

and

$$h_2 - h_1 = c_P (T_2 - T_1)$$

and the ratio of the specific heats, $k = c_P / c_V$. Note: $k = 1.4$ for air at standard temperature and pressure. Usually the volume of the tank, V, is constant, so the mass, m, will vary as the pressure and temperature change.

$$m = \frac{MPV}{RT}$$

For filling an insulated, initially empty vessel, as shown in the previous section, the conservation of energy equation simplifies to: $u_2 = h_{in}$. For an ideal gas with constant specific heats, then $u_2 = c_V T_2$ and $h_{in} = c_P T_{in}$. Combining these gives the final temperature in the tank as:

$$T_2 = \frac{c_P}{c_V}T_{in} = kT_{in}$$

For emptying a tank filled with an ideal gas, we can reconsider Equation 8.12 and apply the assumption of an insulated tank so that Q = 0.

$$\frac{d(mu)_{CV}}{dt} = -\dot{m}_{out}h_{out}$$

Applying the assumptions of an ideal gas with constant specific heats we can write:

$$\frac{d(mc_VT)_{CV}}{dt} = -\dot{m}_{out}c_PT$$

where we are assuming the temperature of the outflow is the same at the temperature inside the tank at any specific point in time. Since the constant-volume specific heat is assumed to be constant, it can be pulled out of the derivative and divided out from the equation:

$$\frac{d(mT)_{CV}}{dt} = -\dot{m}_{out}kT$$

The derivative can be expanded using the product rule:

$$\frac{d(mT)_{CV}}{dt} = T\frac{dm}{dt} + m\frac{dT}{dt}$$

and the rate of change of mass in the system is related to the outflow mass flow rate:

$$\frac{dm}{dt} = -\dot{m}_{out}$$

Combining theses yields:

$$T\frac{dm}{dt} + m\frac{dT}{dt} = \frac{dm}{dt}kT$$

Which can then be re-arranged into:

$$m\frac{dT}{dt} = \frac{dm}{dt}(k-1)T$$

Using separation of variables we have:

$$\frac{dT}{T} = (k-1)\frac{dm}{m}$$

Integrating this expression yields:

$$\frac{T_2}{T_1} = \left(\frac{m_2}{m_1}\right)^{k-1}$$

This expression can be used to calculate the final temperature in the tank if the final mass (or the total mass exiting the tank) is known. If the pressure in the tank is being monitored instead, then we can make use of the ideal gas law with m = MPV/RT, and since V, M, and R are constant for a rigid tank of fixed composition, then we can substitute for the mass:

$$\frac{T_2}{T_1} = \left(\frac{P_2/T_2}{P_1/T_1}\right)^{k-1}$$

which can be simplified to show the final temperature as a function of the final pressure in the tank:

$$\boxed{\frac{T_2}{T_1} = \left(\frac{P_2}{P_1}\right)^{(k-1)/k}}$$

Example 21:
A rigid insulated tank of volume 100 L is initially filed with air at P = 60 bar and T = 300 K. A valve is opened and the tank is discharged to a final pressure of P = 2 bar. Find the final temperature of the air in the tank and the mass of the air left in the tank.

SOLUTION: The pressure is low enough that the ideal gas law can still be used, so that the initial mass in the tank is calculated by:

$$m = \frac{MPV}{RT} = \frac{(29\ kg/kmol)(60\times100{,}000\ Pa)(0.1\ m^3)}{(8314\ \frac{J}{kmol\ K})(300\ K)} = 6.98\ kg$$

The final temperature is calculated using the ideal gas relationship between P&T for an isentropic process:

$$\frac{T_2}{T_1} = \left(\frac{P_2}{P_1}\right)^{(k-1)/k} = \left(\frac{2\ bar}{60\ bar}\right)^{(1.4-1)/1.4} = \left(\frac{1}{30}\right)^{2/7} = 0.378$$

The final temperature is the $T_2 = 0.378\, T_1 = 0.378(300\text{ K}) = 113\text{ K}$, or -160 °C. The final mass is calculated using the ideal gas law, expressed in terms of mass rather than moles:

$$m = \frac{MPV}{RT} = \frac{(29\text{ kg/kmol})(200{,}000\text{ Pa})(0.1 m^2)}{\left(8314\,\frac{J}{kmol\,K}\right)(113\text{ K})} = 0.62\text{ kg}$$

Example 22:
Consider an initially empty scuba tank of volume 11.0 L being filled with air from a supply at a pressure of 200 bar and 300 K. What is the maximum possible temperature of the air in the tank after being filled? What will be the mass of air in the tank?

SOLUTION: Using the equation for an insulated tank (a good question to ask yourself is how do you know this is the right equation to use?), the final temperature would be:

$$T_2 = kT_{in} = 1.4(300\text{ K}) = 420\text{ K}$$

This is about 147 °C. If the tank were not insulated then there would be some heat loss to the environment and the final temperature would be lower. Assuming the final pressure in the tank is equal to the supply pressure, the mass in the tank can be calculated using the ideal gas law:

$$m = \frac{MPV}{RT} = \frac{(29\text{ kg/kmol})(200\times 100{,}000\text{ Pa})(0.011 m^2)}{\left(8314\,\frac{J}{kmol\,K}\right)(420\text{ K})} = 1.83\text{ kg}$$

SIMPLIED FORMS OF ENERGY EQUATION

This handout shows how the general form of the First Law of Thermodynamics to can be simplified for specific applications.

For <u>open</u> systems, we must also consider the energy associated with the fluids (liquids or gases) flowing in and out of the control volume as well as any other energy transfers. It has been assumed that all energy fluxes crossing the control system boundary can be represented either as heat transfers, Q, or mechanical work terms, W.

$$\frac{dE_{CV}}{dt} = \sum \pm \dot{Q} + \sum \pm \dot{W} + \sum_{in} \dot{m}\left(h + \frac{V^2}{2} + gz\right) - \sum_{out} \dot{m}\left(h + \frac{V^2}{2} + gz\right)$$

It is traditional in thermodynamics to define heat transfers as positive if they add energy to the system and negative if they take energy out of the system, and to define the work term as positive if work is done by the system, that is the work takes energy out of the system, and negative if work is absorbed by the system. This convention is arbitrary and is really only useful for heat engines. In this handout the heat transfer and work terms are prefaced with a +/- symbol and it is up to the student to determine whether each term adds energy to the system or takes away energy.

In almost all cases, except for hydro-electric power plants and other hydraulic applications, the potential energy term $g\Delta z$ can be neglected because it is small compared to the other terms.

$$\frac{dE_{CV}}{dt} = \sum \pm \dot{Q} + \sum \pm \dot{W} + \sum_{in} \dot{m}\left(h + \frac{V^2}{2}\right) - \sum_{out} \dot{m}\left(h + \frac{V^2}{2}\right)$$

The vast majority of the problems in undergraduate thermodynamics courses are usually steady state problems. For steady state systems dE/dt = 0. This assumption can not be used for the initial transient startup and cool down phases of operation for devices that normally operate in a steady state, such as power plant boilers and turbines or for filling and emptying problems.

$$0 = \sum \pm \dot{Q} + \sum \pm \dot{W} + \sum_{in} \dot{m}\left(h + \frac{V^2}{2}\right) - \sum_{out} \dot{m}\left(h + \frac{V^2}{2}\right)$$

Many practical devices, such as compressors and pumps and simple turbines, have only one inlet and one outlet. For such devices under the steady state assumption the mass flow rate in equals the mass flow rate out, because there is no net change in mass within the control volume, so the energy equation can be simplified by setting $m_{in} = m_{out}$ and dividing by the mass flow rate of the system:

$$\boxed{\frac{\sum \pm \dot{Q}}{\dot{m}} + \frac{\sum \pm \dot{W}}{\dot{m}} + \left(h + \frac{V^2}{2}\right)_{in} - \left(h + \frac{V^2}{2}\right)_{out} = 0}$$

No further simplifications can be made at this point without selecting a specific application.

TURBINE – Generates mechanical (rotating shaft) power from enthalpy of flowing fluid

They are generally assumed to be well-insulated so there is no heat loss (adiabatic). The only work term is the power generated by the spinning turbine blades. It is also often assumed that the inlet and exit velocities are the same.

$$\boxed{\frac{\dot{W}_{turbine}}{\dot{m}} = h_{in} - h_{out}}$$

If high accuracy calculations are required, the assumption of constant velocity should be double-checked. The assumption of no heat loss may also not be accurate for real turbines.

COMPRESSOR – Takes in mechanical power to compress a gas to higher pressure

Also usually assumed to be insulated from heat transfer losses, and the only power term is the power required to run the compressor.

$$\boxed{\frac{\dot{W}_{compressor}}{\dot{m}} = h_{out} - h_{in}}$$

PUMP – Takes in mechanical power to compress a liquid to higher pressure

$$\boxed{\frac{\dot{W}_{pump}}{\dot{m}} = h_{out} - h_{in}}$$

One note on pumps: If we model the liquid as being incompressible, which is usually a good assumption for reasonable ranges of pressure rise, then the change of enthalpy term can be simplified to $\Delta h = h_{out} - h_{in} = (u + pv)_{out} - (u + pv)_{in} = \Delta u + v\, \Delta p = \Delta u + \Delta p / \rho$.

NOZZLE – Accelerates fluid velocity at the expense of enthalpy

There are no moving parts in a nozzle so no mechanical work can be done. It is also assumed that there are no heat transfer losses.

$$\boxed{h_{in} + \frac{V_{in}^2}{2} - h_{out} - \frac{V_{out}^2}{2} = 0}$$

DIFFUSER – Reduces fluid velocity

There are no moving parts in a diffuser so no mechanical work can be done. It is also assumed that there are no heat transfer losses.

$$h_{in} + \frac{V_{in}^2}{2} - h_{out} - \frac{V_{out}^2}{2} = 0$$

Note: Sometimes it is assumed that the inlet velocity to a nozzle or the outlet velocity from a diffuser is negligible in working problems. Of course these velocities cannot be exactly zero or else nothing would happen in the device. Rather the assumption is that if one velocity is small compared to another, then its kinetic energy per mass, V^2, is really small and can be neglected in the energy equation. If you make this assumption you should double-check its accuracy when you finish the problem.

BOILER – Uses heat transfer to increase enthalpy of working fluid

There are no work terms and kinetic energy changes are neglected.

$$\frac{\dot{Q}_{in}}{\dot{m}} = h_{out} - h_{in}$$

HEAT EXCHANGER – Transfers thermal energy from one fluid stream to another

Heat exchangers are different from all the other devices discussed in this handout in that there are multiple inlets and outlets of fluid flow. It is assumed that the outer surface of the heat exchanger is insulated so that there are no losses to the surroundings.

$$\sum h_{in} = \sum h_{out}$$

One note that should be made for all applications where the kinetic energy of fluid flow is important: We have assumed that the mean flow velocity is the appropriate velocity to use in the kinetic energy term V^2. Strictly speaking, this is only valid mathematically if the flow velocity is uniform across the channel in which the fluid is moving. In reality friction as the solid boundary walls causes the fluid velocity to be lower near the edges of the flow and higher in the middle. Since the kinetic energy term is proportional to the velocity squared, then the total kinetic energy of the flow would be greater than what is calculated by taking the mean velocity squared. For turbulent flows this error in computing the kinetic energy is less than 10% and is usually neglected. For laminar flows in a round pipe this error is 50% due to the parabolic velocity profile.

PSYCHROMETRICS

PHYCHROMETRICS is the study of mixtures of dry air and water vapor. Such a mixture is sometimes referred to as moist air.

ABSOLUTE HUMIDITY (ω) – The ratio of water vapor to dry air on a mass basis

$$\omega = \frac{m_{vapor}}{m_{air}}$$

$$y_{vapor} = \frac{\omega}{\omega + 1}$$

$$\omega = \frac{M_{vapor}}{M_{air}} \frac{x_{vapor}}{x_{air}} = \frac{M_{vapor}}{M_{air}} \frac{P_{vapor}}{P_{air}}$$

Note: M_{vapor}/M_{air} = 18.02 / 28.97 = 0.622 (to 3 significant digits) and P_{air} = P - P_{vapor}.

RELATIVE HUMIDITY (ϕ) – The ratio of the absolute humidity to the maximum possible absolute humidity (at saturation). $0 \leq \phi \leq 1$.

$$\phi = \frac{m_{vapor}}{m_{vapor,saturated}}$$

also

$$\phi = \frac{P_{vapor}}{P_{saturation}(T)}$$

SATURATION – The maximum amount of water vapor that air can hold under particular conditions. If any more water vapor is added it will condense out. The saturation point depends of the temperature (or pressure) of the air. The saturation pressure, $P_{vapor,max} = P_{sat}(T)$ is a function of temperature only and is the maximum possible partial pressure of water vapor in dry air at that temperature.

DRY BULB, WET BULB, and DEWPOINT TEMPERATURES – For a fixed atmospheric or process pressure, water will condense into the liquid phase when the temperature drops below the dew point temperature. The dry bulb temperature is just the regularly measured temperature. The wet bulb temperature is temperature measured in the air when liquid water evaporates to achieve equilibrium with its surroundings and provides a cooling effect.

PROPERTIES OF PSYCHROMETRIC MIXTURES – Treat as an ideal gas mixture. Applies to: Enthalpy, specific heats, density, internal energy and entropy. Equation for enthalpy shown:

$$h = \sum_i y_i h_u = \frac{1}{\omega+1} h_{air} + \frac{\omega}{\omega+1} h_{vapor}$$

RELATIONSHIP BETWEEN ABSOLUTE AND RELATIVE HUMIDITY –

$$\omega = \frac{0.622 \phi P_{sat}(T)}{P - \phi P_{sat}(T)}$$

$$\phi = \frac{\omega P}{(0.622 + \omega) P_{sat}(T)}$$

A *hygrometer* is a device used to measure the humidity. Traditionally, it consists of two thermometers, one of which is attached to or covered with a wicking fabric soaked with water. The two thermometers can then be spun through the air, often with a hand-operated spinning device, which causes the water on the wick to evaporate. This endothermic process will lower the temperature of the wet-bulb thermometer. The other thermometer is the dry bulb thermometer. Any two of three values - dry-bulb temperature, wet-bulb temperature, and relative humidity - can be used to compute the remaining unknown.

Example 23:
Outside air at 5 °C and 60% relative humidity is brought into a building furnace and heated to 25 °C. Find the relative humidity of the air coming out of the furnace if no water is added, and find the rate of heat transfer (in kW) the furnace adds to the air if the volume flow rate at the furnace inlet is 60 m³/min. The pressure is constant at 1 bar.

SOLUTION: First, calculate the absolute humidity of air at the inlet:

$$\omega = \frac{0.622 \phi P_{sat}(T)}{P - \phi P_{sat}(T)}$$

The saturation pressure at 5 °C can be found from sources such as:
https://www.efunda.com/materials/water/steamtable_sat.cfm
http://www.thermopedia.com/content/1150/
We find that: P_{sat}(5 °C) = 0.0087 bar and P_{sat}(25 °C) = 0.0317 bar (we will need this soon).

$$\omega = \frac{0.622\phi P_{sat}(T)}{P - \phi P_{sat}(T)} = \frac{0.622(0.60) * 0.0087 \, bar}{1.0 \, bar - (0.60)*0.0087 bar} = \frac{0.0032 \, bar}{0.995 \, bar} = 0.0032$$

Since no mass transfer occurs in the cooler, the mass of water and air will be constant across the device, so the absolute humidity at the outlet will be the same as that in the inlet: $\omega_1 = \omega_2$. We can find the relative humidity at the outlet by:

$$\phi = \frac{\omega P}{(0.622 + \omega)P_{sat}(T)} = \frac{(0.0032)(1.0 \, bar)}{(0.622 + 0.0032)(0.0317 \, bar)} = 0.16 = 16\%$$

Note the relative humidity has dropped considerably, as warmer air can hold more moisture than cold air. The saturation pressure increases with increasing temperature until the critical point is reached.

To find the rate of heat transfer we need to know the enthalpy of the moist air mixture at both inlet and outlet. Since the temperature change is only 20 degrees, we can assume constant specific heats. But we need an average specific heat value for the air-water mixture. The specific heat of dry air is approximately $c_{P,air}$ = 1.01 kJ/kg-°C, and for water vapor it is $c_{P,w}$ = 1.84 kJ/kg-°C. Thus the specific heat of the moist air is:

$$c_P = c_{P,air} + \omega \, c_{P,w} = 1.01 + (0.0032)(1.84) = 1.016 \text{ kJ/kg-°C}$$

This is less than 1% different than the specific heat of dry air, so we could just use the specific heat of dry air and be accurate enough for most engineering calculations. So does the water vapor matter in the energy balance? If any water had been vaporized or condensed it would be important, as the heat of vaporization of water at 5 °C is 2453.5 kJ/kg.

The change in enthalpy is:

$$\Delta h = c_P \, \Delta T = (1.016 \text{ kJ/kg-°C})(20 \text{ °C}) = 20.32 \text{ kJ/kg}$$

To calculate the energy required from the furnace, we need to know the mass flow rate of air. The volume flow rate of air is known (most flowmeters measure volume flowrate instead of mass flowrate), so we also need the density of the air. To obtain a very precise answer one could calculate the average molecular weight of the moisture-air mixture (water vapor is less dense than air) and then use the ideal gas law to calculate the density of the moist air at the point where the flowrate is measured. I am satisfied with an approximate answer, and will use the value of density for air at standard atmosphere conditions (approximately 1.2 kg/m^3). Thus the required heating is:

$$\dot{Q} = \dot{m}\Delta h = \left(1.2\frac{kg}{m^3}\right)\left(60\frac{m^3}{s}\right)\left(20.32\frac{kJ}{kg}\right) = 1463\ kW$$

Note that using constant specific heats to calculate the change in enthalpy works in this problem because of the small temperature difference and because there was no phase change (water was not being evaporated or condensed). If phase change of the moisture had occurred, then the enthalpy of vaporization, h_{fg}, would have to be accounted for.

My general rule of thumb is that if the temperature change is less than 200 degrees you can usually assume constant specific heats for the degree of accuracy for typical engineering calculations, but for larger temperature differences you need to use tabulated enthalpy values from a suitable database.

What do you think would happen if a steam of cold moist air is mixed with a steam of hot dry air? Would the result be at the average humidity of the two streams being mixed? Let's test this by calculating an example:

Example 24:
a stream of 1 kg/s air at 10 °C and 90% humidity mixing with a stream of air at 1 kg/s and 30 °C and 10% humidity. Will the outlet be at 50% humidity?

SOLUTION: We can answer this question by using conservation of mass to calculate the properties of the mixed stream of moist air. First we need the saturation pressures of the two streams being mixed, so we can calculate the absolute humidity of each.

We find that: $P_{sat}(10\ °C) = 0.01228$ bar and $P_{sat}(30\ °C) = 0.04247$ bar, using the tables online at
https://www.engineeringtoolbox.com/water-properties-d_1573.html
Note many databases and printed steam tables will have pressures in kPa, so you must divide by 100 to get to bars.

$$\omega_1 = \frac{0.622\phi P_{sat}(T)}{P - \phi P_{sat}(T)} = \frac{0.622(0.90)*0.01128\ bar}{1.0\ bar - (0.90)*0.01128\ bar} = \frac{0.00631\ bar}{0.990\ bar} = 0.00637$$

$$\omega_2 = \frac{0.622\phi P_{sat}(T)}{P - \phi P_{sat}(T)} = \frac{0.622(0.10)*0.04247\ bar}{1.0\ bar - (0.10)*0.04247\ bar} = \frac{0.00264\ bar}{0.996\ bar} = 0.00265$$

So by conservation of mass we can calculate the mass of the outlet stream:
$$\dot{m}_1 + \dot{m}_2 = \dot{m}_3$$

So the outlet stream has a mass flow rate of 2 kg/s (you didn't really need an equation to figure that out, did you?). Also the mass of water vapor is conserved. So we need to find out how much water is in each of the two inlet streams.

$$\omega = \frac{m_{vapor}}{m_{air}} \text{ and } \dot{m}_{vapor} + \dot{m}_{air} = \dot{m} = 1 \text{ kg/s}$$

Since ω and m are known, we have two equations and 2 unknowns, so we can solve for the mass of water vapor:

$$\dot{m}_{vapor,1} = \frac{\omega}{\omega + 1}\dot{m} = 0.00633 \text{ kg/s}$$

and

$$\dot{m}_{vapor,2} = \frac{\omega}{\omega + 1}\dot{m} = 0.00264 \text{ kg/s}$$

and the total mass of water vapor in the outlet stream is:

$$\dot{m}_{vapor,3} = \dot{m}_{vapor,1} + \dot{m}_{vapor,2} = 0.00897 \text{ kg/s}$$

and the absolute humidity of the outlet stream is:

$$\omega_3 = \frac{m_{vapor}}{m_{air}} = \frac{m_{vapor}}{m - m_{vapor}} = \frac{0.00897}{2.0 - 0.00897} = 0.00451$$

(You might have notices this is the average of ω_1 and ω_2). We will need to know the temperature of the outlet stream. Conservation of energy for this problem simplifies to:

$$\dot{m}_1 h_1 + \dot{m}_2 h_2 = \dot{m}_3 h_3$$

From the previous problem we know that the specific heat of the moist air will not change significantly. So if we have constant and equal specific heats, along with equal inlet mass flow rates, the energy balance simplifies to:

$$T_3 = \frac{T_1 + T_2}{2}$$

So the temperature of the combined mixed stream will be approximately 20 °C. To convert between absolute and relative humidity, we will need the vapor pressure of water at this temperature. $P_{sat}(20 °C)$ = 0.02339 bar.

$$\phi = \frac{\omega P}{(0.622 + \omega)P_{sat}(T)} = \frac{0.00451(1.0 \text{ bar})}{(0.622 + 0.00451)(0.02339 \text{ bar})} = 0.308 = 31\%$$

So the relative humidity is less than 50%, because the vapor pressure increases with temperature, and the relatively humid cold air cannot hold much moisture content, so when it is mixed with warmer dry air the relative humidity drops quite a bit.

General Two-Phase Mixtures

The previous section on *psychrometrics* dealt with mixtures of gaseous air and water in either liquid or vapor state. We also need tools to deal with multicomponent mixtures in general. Quite often though we find we are interested in a liquid being vaporized into air (such as gasoline vaporizing inside the cylinder of a car engine so that it can burn) or substances being absorbed into liquid water, since water and air are the two most common fluids we deal with in engineering practice.

The two simplest models for two-component, two-phase mixtures are Raoult's Law and Henry's Law.

Raoult's law states that the partial pressure of component i in the gas phase is equal to its mole fraction in the liquid phase times its saturation pressure at the mixture temperature. The vapor pressure/saturation pressure is that of the pure component. Raoult's law works well at low pressures.

$$P_i = x_{i,liq} P_i^{sat}$$

Note that usually in thermodynamics we use the subscript f for the liquid phase, but here *liq* or sometimes L is often used for liquid. For a two-component mixture, i = 1 or 2. The vapor pressure of a solution of a non-volatile solute is equal to the vapor pressure of the pure solvent at that temperature multiplied by its mole fraction. Raoult's Law for a binary mixture of two components labeled A and B is:

$$P_{B,gas} = x_{B,liq} P_B^{sat}$$

So if species A is a liquid solvent (such as water), and species B is a gas, then the mole fraction of dissolved gas in the water in an ideal solution is:

$$x_{B,gas} = \frac{P_{B,gas}}{P_B^{sat}}$$

Note that for substances that are liquids at room temperature the vapor pressure is less than 1 atm, and for substances that are gases at room temperature, the vapor pressure is above 1 atm.

Raoult's Law works well for dilute solutions. That is to say, reasonably accurate results are obtained when the mole fraction of the liquid (A) is near 1.0. When working problems with Raoult's law, you can use tabular data or you can use a curve fit correlation like the Antoine equation to get $P_{vap}(T)$. If the mole fraction of the liquid is not close to 1 or the solution otherwise does not follow an ideal solution, Henry's Law can be a preferable option.

Henry's Law

Henry's law is a correlation that tells us how much of a gas (B) can be dissolved in a liquid (A), usually in the form of:

$$x_{B,gas} = K_H \, P_{B,gas}$$

where K is an empirical Henry's law constant, that must be looked up for a particular liquid-gas system. Henry's Law constants for many substances can be found at the NIST Chemistry Webbook (https://webbook.nist.gov/). Henry's Law constants depend on temperature. Henry's Law is good approximation when the solubility is low, and it does not work well for gases at high pressure. The units for Henry's Law constant depend on the form of the equation being used. Often it specifies the pressure in units of atm. Also, sometimes the concentration of the dissolved gas, which uses units of mol/L, is used instead of the mole fraction.

To summarize, Henry's Law works well at low concentrations and Raoult's Law works well at high concentrations.

Applications
Examples of processes involving two-component gas-liquid mixtures include osmosis, reverse-osmosis separation, evaporation, distillation, drying, humidification, condensation, and dehumidification.

Combustion

What is a combustion reaction? It is an **exothermic** (heat releasing) chemical reaction of the type:
fuel + oxidizer → products

Figure 31: Combustion-powered rocket engines and boosters of space shuttle. From NASA Image Archives.

Combustion is conversion of chemical energy to thermal energy. Reactions usually take place in the gas phase, with the exception of certain solid fuel systems. The oxidizing agent is usually the oxygen in air, since it is free and readily available. The fuel and oxidizer together are collectively referred to as the **reactants** for a combustion reaction. The **products** are the chemical species that result from the combustion reaction. Major products from a combustion reaction typically include carbon dioxide and water vapor. Some of the minor products of combustion have deleterious effects on health and environment in spite of their small concentrations.

The enthalpy of formation is the energy required to make or break the bonds of a molecule. It is the chemical bond energy. The total enthalpy of a substance is defined as the sum of the chemical enthalpy and the sensible enthalpy. The sensible enthalpy is the energy associated with the thermal state of the material, which is the concept that most first semester thermodynamics courses are concerned with.

$$h_i(T) = h°_{f,i}(T_{ref}) + \Delta h_{s,i}(T-T_{ref})$$

T_{ref} is usually taken to be 25 °C = 298.15 K and the corresponding P_{ref} is 1 atm = 101,325 Pa. The heat of formation for an element in its naturally occurring state is zero. For example, at STP (standard temperature and pressure) air is in its naturally occurring state of a gas, so the heat of formations for gaseous O_2 and N_2 at 298 K are 0. Table 19 shows the chemical enthalpies for chemical species commonly encountered in the combustion of hydrocarbon fuels.

Table 19: Chemical enthalpies for commonly encountered substances in combustion reactions. [JANAF71]

Name	Chemical Formula	Enthalpy of formation at 298 K (kJ/kmol)
Carbon (solid)	C	0
Methane (gas)	CH_4	-74,870
Hydrogen (gas)	H_2	0
Oxygen (gas)	O_2	0
Carbon Monoxide	CO	-110,530
Carbon Dioxide	CO_2	-393,520
Water Vapor	H_2O	-241,820
Water (liquid)	H_2O	-285,820

For a given fuel, we want to find the maximum possible energy that can be released by burning the fuel. This is termed as the *heat of combustion* for that fuel, also referred to as the *heating value* or *enthalpy of combustion* of the fuel. The maximum heat release occurs for hydrocarbon fuels when all the carbon in the fuel is completely oxidized to carbon dioxide and all the hydrogen in the fuel is completely oxidized to water, so that the products of combustion are in the lowest possible energy state. The heat of combustion for a chemical reaction is defined as the net change in chemical energy from reactants to products:

$$\Delta H_{comb} = \sum_{Reactants} N_i h^0_{f,i} - \sum_{Products} N_i h^0_{f,i}$$

where N_i is the number of moles of each species in the balanced reaction, based on one mole of fuel.

Example 25:
Calculate the heat of combustion for methane (CH_4).

SOLUTION: First we must write the balanced chemical reaction for methane combusting with oxygen, which is:

$$CH_4 + 2\ O_2 \rightarrow CO_2 + 2\ H_2O$$

From Table 6.1, for methane the heat of formation is -74,870 kJ/kmol, for oxygen is 0, for carbon dioxide is -393,520 kJ/kmol, and for water vapor is -241,820 kJ/kmol. (The molecular weight of methane is 1*12 + 4*1 = 16 kg/kmol, so on a mass basis the heat of formation is -74,870 kJ/kmol / 16.04 kg/kmol = -4668 kJ/kg.) To calculate the heat of combustion using Eq. 6.2 we write:

$$\Delta H = h^0_{f,CH4} + 2h^0_{f,O2} - h^0_{f,CO2} - 2h^0_{f,H2O}$$

Substituting in the numerical values:

$$\Delta H = -74{,}870 \text{ kJ} + 0 \text{ kJ} - (-393{,}520 \text{ kJ}) - 2 \times (-241{,}820 \text{ kJ}) = 802{,}290 \text{ kJ}$$

Thus, the heat of combustion is ΔH = 802,290 kJ per kmol of CH_4. Since heating values are more commonly reported on a mass basis, we can convert this using the molecular weight of methane. M = 16.04 kg/kmol.

$$\Delta H = 802{,}290 \text{ kJ/kmol} / 16 \text{ kg/kmol} = 50{,}018 \text{ kJ/kg}$$

The value of the heat of combustion calculated in Example 25 is referred to as the **Lower Heating Value** (LHV). The LHV is applicable when the water in the combustion products is in the gaseous state. The enthalpy of formation for water vapor is -241,820 kJ/kmol, while for liquid water it is -285,820 kJ/kmol. If the water in the products can be condensed to the liquid state, then more heat can be released, and the resulting heat of combustion is referred to as the **Higher Heating Value** (HHV). So for the specific example of methane, if we replace the water vapor in the products with liquid water, then the heat of combustion is:

$$\Delta H = -78{,}870 \text{ kJ} + 0 \text{ kJ} - (-393{,}520 \text{ kJ}) - 2 \times (-285{,}820 \text{ kJ}) = 886{,}290 \text{ kJ}$$

Which on a per mass basis is ΔH = 886,290 kJ/kmol / 16.04 kg/kmol = 55,255 kJ/kg. Thus for methane the higher heating value is 10.5% larger than the lower heating value.

Table 20: Properties of common combustion fuels. Data from [DOE94], [NACA56], [NIST08]. Heating value in megajoules. 1 MJ = 1000 kJ.

Fuel	Chemical Formula	Molecular Weight (kg/kmol)	Stoichiometric A/F ratio	Lower Hearing Value (MJ/kg)
Hydrogen	H_2	2.02	34.3	120.0
Methane	CH_4	16.04	17.2	50.0
Propane	C_3H_8	44.1	15.7	46.3
Methanol	CH_3OH	32.04	6.5	20.1
Ethanol	C_2H_5OH	46.07	9.0	27.0
Gasoline	mixture	~102	14.7	42.4
Diesel Fuel	mixture	~200	14.7	42.6

Table 21: Properties of coals. Data from [Deskbook].

Rank	Moisture Content	Fixed Carbon	Ash Content	Heating Value (kJ/kg)
Anthracite	3-6%	75-85%	4-15%	28,000-31,500
Bituminous	2-15%	50-70%	4-18%	28,000-33,500
Subbituminous	10-25%	31-55%	3-12%	18,500-25,500
Lignite	35-45%	25-30%	4-15%	14,000-17,500

REFERENCES

[ANL10] J. Sullivan and L. Gaines. A Review of Battery Life-Cycle Analysis: State of Knowledge and Critical Needs. ANL/ESD/10-7. 2010.

[Baxter06] R. Baxter. Energy Storage: A Nontechnical Guide. PennWell. 2006.

[Deskbook] S. Gladstone. Energy Deskbook. U.S. Dept. of Energy. 1982.

[DOE92] DOE Fundamentals Handbook: Thermodynamics, Heat Transfer, and Fluid Flow. Vol. 1. DOE-HDBK-1012/1-92. 1992.

[DOE94] Alternatives to Traditional Transportation Fuels. DOE/EIA-0585/0. 1994.

[DOE99] Market-Based Advanced Coal Power Systems. DOE/FE-0400. 1999.

[DOE03] Improving Compressed Air System Performance. U.S. Department of Energy, Energy Efficiency and Renewable Energy (EERE), 2003.

[EERE04] Energy Tips – Steam. DOE/GO-102004-2006 Sept.2004 Steam Tip Sheet #22

[EERE11] R. Wiser & M. Bolinder. 2010 Wind Technologies Market Report. U.S. Dept. of Energy, Energy Efficiency & Renewable Energy, 2011.

[EPRI03] EPRI-DOE Handbook of Energy Storage for Transmission & Distribution Applications, EPRI, Palo Alto, CA, and the U.S. Department of Energy, 2003.

[FAA] FAA. Pilot's Handbook of Aeronautical Knowledge. 2003. FAA-H-8083-25.

[FERC06] Taum Sauk Pumped Storage Project (No. P-2277), Dam Breach Incident FERC Independent Panel of Consultants (IPOC) Report, May 25, 2006

[Hirschfelder54] J. Hirschfelder, C. Curtiss, and R. Bird. Molecular Theory of Gases and Liquids. Wiley. 1954.

[Hydro96] Hydroelectric Pumped Storage Technology International Experience. American Society of Civil Engineers. 1996.

[JANAF71] JANAF Thermochemical Tables. D. Stull & H. Prophet. NSRDS NBS 37. 1971.

[NACA56] R. Hibbard. Evaluation of Liquefied Hydrocarbon Gases as Turbojet Fuels. NACA-RM-E56I21. 1956.

[NASA69] L. Nichols. Comparison of Brayton and Rankine Cycle Magnetogasdynamic Space-Power Generation Systems. NASA TN D-5085. 1969.

[NBS74] NBS Technical Note 361. Liquid Densities of Oxygen, Nitrogen, Argon and Parahydrogen. H. Roder. 1974.

[NBS81] NBS Monograph 168. Selected Properties of Hydrogen. R. McCarty, J. Hord. H. Roder. 1981.

[NIST95] Theoretical Evaluation of the Vapor Compression Cycle With a Liquid-Line/ Suction-Line Heat Exchanger, Economizer, and Ejector. NISTIR 5606. 1995.

[NIST08] NIST Chemistry WebBook, NIST Standard Reference Database Number 69, Eds. P.J. Linstrom and W.G. Mallard, National Institute of Standards and Technology.

[NREL02] M. O'Keefe and K. Vertin. An Analysis of Hybrid Electric Propulsion Systems for Transit Buses. NREL/MP-540-32858. 2002.

[PSH10] Pumped Storage Hydropower. Summary Report. Oak Ridge National Laboratory. 2010.

[Tipsheet04] Compressed Air Tip Sheet #1. U.S. DOE EERE. Aug. 2004.

THERMODYNAMICS EQUATION SHEET

First Law: $E = m(u + v^2/2 + gz)$ enthalpy $h = u + pv = u + p/\rho$

Closed system: $\Delta E = Q - W$

Open system: $\frac{dE}{dt} = \sum \pm \dot{Q} + \sum \pm \dot{W} + \sum_{in} \dot{m}\left(h + \frac{v^2}{2} + gz\right) - \sum_{out} \dot{m}\left(h + \frac{v^2}{2} + gz\right)$

Second Law:

     Closed system: $S_2 - S_1 = \int \frac{1}{T_b} dQ + \sigma$

     Open system: $\frac{dS}{dt} = \sum \pm \frac{\dot{Q}}{T} + \sum_{in} \dot{m}s - \sum_{out} \dot{m}s + \dot{\sigma}$

Ideal Gas Law: $PV = NR_uT$ $m = N*M$ $v = V/m$ $\rho = PM/R_uT$
               $R_U = 8314$ J/kmol-K $R = R_u / M$

Specific Heats: $c_P = dh/dT$ $c_V = du/dT$ $R = c_P - c_V$ $k = c_P / c_V$

Phase change: $x = m_{vapor} / m_{total}$ $u_{mix} = (1-x)*u_f + x*u_g$ (also true for h, v, s)

Boundary Work: $W = \int P \, dV$
     Isothermal process of ideal gas: $PV = C$
     Isentropic process of ideal gas: $PV^k = C$

Conservation of mass: $\frac{dm_{CV}}{dt} = \sum \dot{m}_{in} - \sum \dot{m}_{out}$ $\dot{m} = \rho A V$

Isentropic efficiencies: $\eta_T = W_{act}/W_{isen}$ $\eta_C = W_{isen}/W_{act}$ $\eta_N = V^2/V_s^2$

Isentropic process of an ideal gas: $\frac{P_2}{P_1} = \left(\frac{V_1}{V_2}\right)^k$ $\frac{T_2}{T_1} = \left(\frac{V_1}{V_2}\right)^{k-1}$ $\frac{T_2}{T_1} = \left(\frac{P_2}{P_1}\right)^{(k-1)/k}$

Carnot cycles: $Q_H / Q_L = T_H / T_L$ $\eta = 1 - T_L / T_H$ for Carnot power cycle

Coefficient of Performance: Refrigerator $\beta = Q_C / W_{NET}$ Heat pump $\gamma = Q_H / W_{NET}$

Availability: $X = (E - U_o) + P(v - v_o) - T_o(S - S_o)$

Flow exergy: $\psi = (h - h_o) + T_o(s - s_o) + V^2/2 + gz$

Properties of air @ STP: $k = 1.4$ $M = 29$ kg/kmol $R = 287$ J/kg-K
     $c_P = 1005$ J/kg-K $c_V = 718$ J/kg-K

Properties of water @ STP: $k = 1.0$ $M = 18$ kg/kmol $\rho = 1000$ kg/m^3
     $c_P = 4184$ J/kg-K $c_V = 4184$ J/kg-K

$$\eta_{Otto} = 1 - \frac{1}{CR^{k-1}} \qquad \eta_{Brayton} = 1 - \frac{1}{PR^{(k-1)/k}}$$

Gibbs free energy: $\quad g = h - T s \quad$ Helmholtz free energy: $a = u - T s$

Mass fraction: $y_i = \frac{m_i}{m}$ and mole fraction: $x_i = \frac{N_i}{N}$ $\quad \sum y_i = 1$ and $\sum x_i = 1$

Mixture mean molecular weight: and $\bar{M} = \sum_i x_i M_i \qquad y_i = x_i \frac{M_i}{\bar{M}}$

Psychrometrics: $M_{vapor}/M_{air} = 18.02 / 28.97 = 0.622$ (to 3 significant digits) and $P_{air} = P - P_{vapor}$

Absolute humidity: $\omega = \frac{m_{vapor}}{m_{air}}$ and $\omega = \frac{M_{vapor}}{M_{air}} \frac{x_{vapor}}{x_{air}} = \frac{M_{vapor}}{M_{air}} \frac{P_{vapor}}{P_{air}} = \frac{0.622 P_{vapor}}{P - P_{vapor}}$

Relative humidity: $\phi = \frac{m_{vapor}}{m_{vapor,saturated}}$ also $\phi = \frac{P_{vapor}}{P_{saturation}(T)}$

$\phi = \frac{\omega P}{(0.622 + \omega) P_{sat}(T)}$ and $\omega = \frac{0.622 \phi P_{sat}(T)}{P - \phi P_{sat}(T)}$

$y_{vapor} = \frac{\omega}{\omega + 1} \qquad h = \sum_i y_i h_u = \frac{1}{\omega + 1} h_{air} + \frac{\omega}{\omega + 1} h_{vapor}$

enthalpy per mass of dry air defined as: $\quad h = h_{air} + \omega\, h_{vapor}$
For -10 °C < T < 50 °C $\qquad h_{vapor} \approx 2500.9 + 1.82\, T$ [kJ/kg]
(thus $c_{P,VAPOR} \approx 1.82$ kJ/kg-K for a very limited range of temperature)

Combustion: For air, for every 1 kmol of O_2 there is 3.76 kmol of N_2

Heats of formation at reference state of 298 K and 1 atm (101.3 kPa)

	H_2	O_2	N_2	H_2O	CO	CO_2
Δh^0_f (kJ/kmol)	0	0	0	-241,800	-110,500	-393,500
c_P (kJ/kmol-K)	28.87	29.32	29.07	33.45	29.07	37.20

Lower heating values (kJ/kg) at reference state of 298 K and 1 atm (101.3 kPa)

	CH_4	C_2H_6	C_3H_8	C_4H_{10}	C_5H_{12}	C_8H_{18}
Δh^0_f (kJ/kmol)	-74,830	-84,670	-103,850	-124,730	-146,440	-208,450
LHV (kJ/kg)	50,020	47,490	46,360	45,740	45,360	44,790

Atomic weights: $\quad$ H – 1, C – 12, N – 14, O – 16 [g/mol or kg/kmol]
1 gmol of a substance contains N_A molecules of that substance
$N_A = 6.02 \times 10^{23}$ molecules/gmol or $= 6.02 \times 10^{26}$ molecules/kmol

Dalton's law of partial pressures: $P = \sum P_i \qquad$ For an ideal gas $P_i = P (N_i / N)$
Pressure 1 bar $= 10^5$ Pa $= 100,000$ Pa $= 100$ kPa $= 0.1$ MPa

Useful Websites

Khan Academy Video Lessons: (if you have the print version of this book, it will be easier to go to the Khan Academy website and search for the titles of the videos you are interested in). Most videos are 20 minutes or less

Ideal Gas Law –
http://www.khanacademy.org/science/chemistry/ideal-gas-laws/v/ideal-gas-equation-example-3
PV diagrams –
http://www.khanacademy.org/science/physics/thermodynamics/v/pv-diagrams-and-expansion-work
Carnot Cycle –
http://www.khanacademy.org/science/physics/thermodynamics/v/carnot-cycle-and-carnot-engine
Engine Efficiency –
http://www.khanacademy.org/science/physics/thermodynamics/v/efficiency-of-a-carnot-engine
Combustion reaction –
http://www.khanacademy.org/science/physics/thermodynamics/v/stoichiometry-example-problem-2
Stoichiometry –
http://www.khanacademy.org/science/physics/thermodynamics/v/empirical-and-molecular-formulas-from-stoichiometry
PV diagrams -
http://www.khanacademy.org/science/physics/thermodynamics/v/pv-diagrams-and-expansion-work

How Stuff Works – Good explanations of basic machinery, usually with diagrams and animations
Refrigerators -
http://www.howstuffworks.com/refrigerator.htm/printable
Gasoline Engines -
http://auto.howstuffworks.com/engine.htm/printable
Diesel Engies -
http://www.howstuffworks.com/diesel.htm/printable
Heat Pumps -
https://home.howstuffworks.com/home-improvement/heating-and-cooling/heat-pump.htm/printable
Gas Turbines -
http://science.howstuffworks.com/transport/flight/modern/turbine.htm/printable
Sump Pump -
https://home.howstuffworks.com/home-improvement/plumbing/sump-pump.htm/printable
Air Conditioners -
https://www.techstuffpodcast.com/podcasts/air-conditioners.htm
Nuclear Power Plants -
https://www.techstuffpodcast.com/podcasts/how-nuclear-power-plants-work.htm
Hydroelectric Power Plants -
https://science.howstuffworks.com/environmental/energy/hydropower-plant.htm/printable
Solar Thermal Power -
https://science.howstuffworks.com/environmental/green-tech/energy-production/solar-thermal-power.htm/printable

Miscellaneous -

- http://www.energyquest.ca.gov/how_it_works/refrigerator.html
- http://www.sfsb.unios.hr/test/testhome/vtAnimations/animations/chapter09/refrigeration/index1.html
- http://www.gunt.de/download/thermodynamics%20of%20refrigeration_english.pdf
- http://www.geaviation.com/education/engines101/
- http://www.explainthatstuff.com/diesel-engines.html
- http://world.honda.com/powerproducts-technology/exlink/
- http://www.animatedengines.com
- DOE Energy Information Administration (EIA) http://www.eia.gov
- National Renewable Energy Lab (NREL) – http://www.nrel.gov
- http://hyperphysics.phy-astr.gsu.edu/hbase/heacon.html#heacon

Thermodynamic Property Data - Online databases include the following:

- NIST Chemistry WebBook - http://webbook.nist.gov/chemistry/
- IAPWS - http://www.steamtablesonline.com/steam97web.aspx - water property calculator
- Steam calculator - also does some calculations for power plant components https://www4.eere.energy.gov/manufacturing/tech_deployment/amo_steam_tool/propSaturated?random=Temperature
- Kaye & Laby Tables of Physical & Chemical Constants - http://www.kayelaby.npl.co.uk/
- Cantera - http://code.google.com/p/cantera/ - open source code for chemical equilibrium and kinetics calculations
- HOT open source thermodynamic tools - http://hot-tdb.sourceforge.net/

Recommended Books – You may be able to find in a library, or to find used copies cheaply:

- Entropy Demystified by A. Ben-Naim
- CRC Handbook of Chemistry and Physics
- Perry's Chemical Engineers Handbook
- Chemical Properties Handbook by C. Yaws
- ASHRAE Handbook – Fundamentals (Refrigerant properties)
- Thermochemical Data of Pure Substances by I. Barin.
- Small Things Considered: Why There Is No Perfect Design by Henry Petroski
- To Engineer is Human: The Role of Failure in Successful Design by Henry Petroski
- When Technology Fails by Neil Schlager

YouTube videos:

Property charts for H_2O: https://www.youtube.com/watch?v=SFjNByAz03w&feature=youtu.be
Organic Rankine Cycle – http://www.youtube.com/watch?v=k3xkGCaFjdw
Geothermal power with refrigerant loop - http://www.youtube.com/watch?v=Uv8bTAGr0tU
Heat Pumps - http://www.youtube.com/watch?v=14MmsNPtn6U
Coal power plant - http://www.youtube.com/watch?v=e_CcrgKLyzc

Note to Instructors

What is the most important thing your students can take out of this course?
What is going to be most useful to them in their careers as practicing engineers?

I suggest the most important thing is **not** memorizing a textbook definition of entropy or the various statements of the second law of thermodynamics. For engineers, the most important thing is learning how to solve problems, including how to find the information they need to solve those problems and developing the persistence to overcome life's challenges.

My own experiences in teaching thermodynamics (and other courses) have led me to try many pedagogical experiments. If something is not working, then why not try something different? Not all of these experiments were successful. My experiences in ABET accreditation helped me answer the question, "How would you know if what you are doing in your class is working?"

This led to my experiments with Standards-Based Grading (some prefer the term Criteria-Based Grading). I would give partial credit on quizzes and exams for incorrect answers, and when I calculated the average scores on quizzes and exams in my EXCEL spreadsheet, it appeared that the class was doing okay. But when I had to go back through the marked assignments to find what percentage of my students had completed the assigned learning objective for ABET, I found very few of them were actually able to complete the objective correctly. The ideas that came out of this experience are summarized in my paper: Post, S. (2017). Standards-based grading in a thermodynamics course. *International Journal of Engineering Pedagogy*, 7(1).

I also wonder if much of the standard way we teaching engineering lectures is inefficient. I would often start with some theory in class, give a mathematical derivation to arrive at a particular equation, and then show how that equation could be used to solve an example problem, and then assign the students to work a similar problem. I suppose that is slightly better than just lecturing straight through a whole hour every day and telling the students, "trust me, you will need to know this later." But I think project-based learning is a much better way to motivate students, puts them in charge of their own education, and also provides avenues for them to explore learning opportunities I would never of thought of myself. In summary, I wonder if presenting new material would be more effective if application came before theory, and the focus was kept on practical devices (rather than theoretical abstractions like heat reservoirs). This in turn will lead the students to assess what processes and concepts need to be understood. At least that is my theory.

As teaching thermodynamics relates to this particular book - Why assign a 500+ page textbook that your students aren't going to read? And why cover material in your class just because it is in the book? I think a low-cost text around 100 pages that just covers the basics is more effective and more accessible to your students. And of course if you think I have omitted something critical, you can easily supplement this book with your own notes, and will have saved your students some money.

Appendices: Data Tables

Using Tabular Data Tables

Your options for finding thermodynamic property data include interactive online resources like the NIST Chemistry Webbook, purchasing commercial software such as EES (your university may have an academic license available), or using tabulated property data tables, either printed in a book or formatted for a webpage. You may also be able to find handbooks of property data in a library, particularly for substances that are not as common. In the near future, all this property information will be available on computers (if it is not already – there are of course phone Apps with steam tables available now), but many universities still do not have facilities available for students to take exams on computers, so you may have to use printed tables for your exams.

Whatever data source you are using, you need to know the following:
- The substance you are using
- Which information (property) you need
- What properties you already know

These parameters will help you determine if you have enough information to solve the problem.

It takes two variables to define a state - We often use pressure and temperature since they are most easily measured, but can be any two from (P, T, u, h, v, s). The only exception is that for ideal gases, the enthalpy, h, and internal energy, u, are functions of temperature (T) only. Inside the vapor dome, pressure and temperature are no longer independent of each other, in fact they are uniquely linked, so another parameter is needed to define the state – that is the *quality*, x. A value of quality x = 0.0 corresponds to pure liquid and x = 1.0 corresponds to pure vapor.

Saturation tables

For typical saturation tables arranged by pressure, the first column is pressure (independent variable) and the second is the saturation (boiling temperature) at each pressure. The next two columns are the specific volume of the pure liquid and gas phases at the saturation condition. Specific volume is the inverse density. In order to make the values fit on the page better, liquid specific volumes are sometimes multiplied by 1000. So to get the actual specific volume take the value in the table and divide by 1000. For example, saturated liquid water at 1.00 bar has a specific volume, v, of 0.0010432 m³/kg, and the density is the inverse of that, or 959 kg/m³. Your book uses the subscript *f* for the liquid phase and the subscript *g* for the gas phase. The next two columns are the internal energy of the liquid and gas (vapor) phases. Note that the internal energy of the vapor is always higher than that of a liquid of the same material at the same temperature. The next three columns are for enthalpy. Enthalpy is used for open flow systems, and internal energy is used in closed systems. The enthalpy of vaporization, h_{fg}, is the energy required to vaporize the saturated liquid. By definition then, $h_{fg} = h_g - h_f$. The final two columns are for the entropy of the liquid and gas phases.

Superheated vapor tables
If you know that the substance you are working with is completely a gas with no chance of being saturated, then you use the single-phase gas tables. These tables are usually arranged in blocks of constant pressure, with data for varying temperature in each block. The first row is the properties at the saturation temperature, which duplicates the data in the saturation table for saturated vapor. All the other rows are for higher temperatures and single-phase super-heated gas. In some cases it may be necessary to interpolate twice to get the desired data.

It may not always be obvious which table you need. You may have to make an assumption and work the problem, and if you do not get a reasonable answer, start over and try again.

Linear Interpolation
To use printed tables you need to know how to perform a *linear interpolation* – most scientific calculators will do this for you (you may need to read the directions).

Example A1:
Find the boiling point of water at 0.84 bar (atmospheric pressure in Denver).

SOLUTION: Recognizing that you need the Saturation table, you can use **Table A5** in the Appendix. This table is arranged by temperature as the independent variable. Some saturation tables are arranged by pressure, but either will do. With Table A5 being arranged by temperature, the second column is the saturation pressure at each temperature. Most printed tables will not have properties of water at 0.84 bar. The closest values in the table to the pressure we need are 0.7018 and 1.0142 bars, as shown in the selection from Table A5 below:

T (°C)	P (bar)
90	0.7018
100	1.0142

So the temperature will be between 90 and 100 °C. To find the exact value we will use interpolation, creating a straight line between the two (x,y) data points in the table. The equation of this line is:

$$\frac{T-90}{100-90} = \frac{P-0.7018}{1.0142-0.7018}$$

With the value of P fixed at P = 0.84 bar, there is one equation and one unknown (T). Solving yields: T = 90 °C + (10 °C)(0.138/0.312) = 94.42 °C (Please limit yourselves to a reasonable number of digits in your answers). For comparison www.steamtablesonline.com gives 94.80 °C.

Vapor Domes

Figure A1 shows a P-v plot for water with the vapor dome drawn. Points to the left of the blue curve represent compressed liquid water (low specific volume, or high density). Points to the right of the red curve represent superheated gas or vapor, with high specific volume and low density. Points above the top of the curve (where the liquid and vapor lines meet at the critical point) are supercritical fluids. Points below the curve are gas-liquid mixtures, at a saturated state.

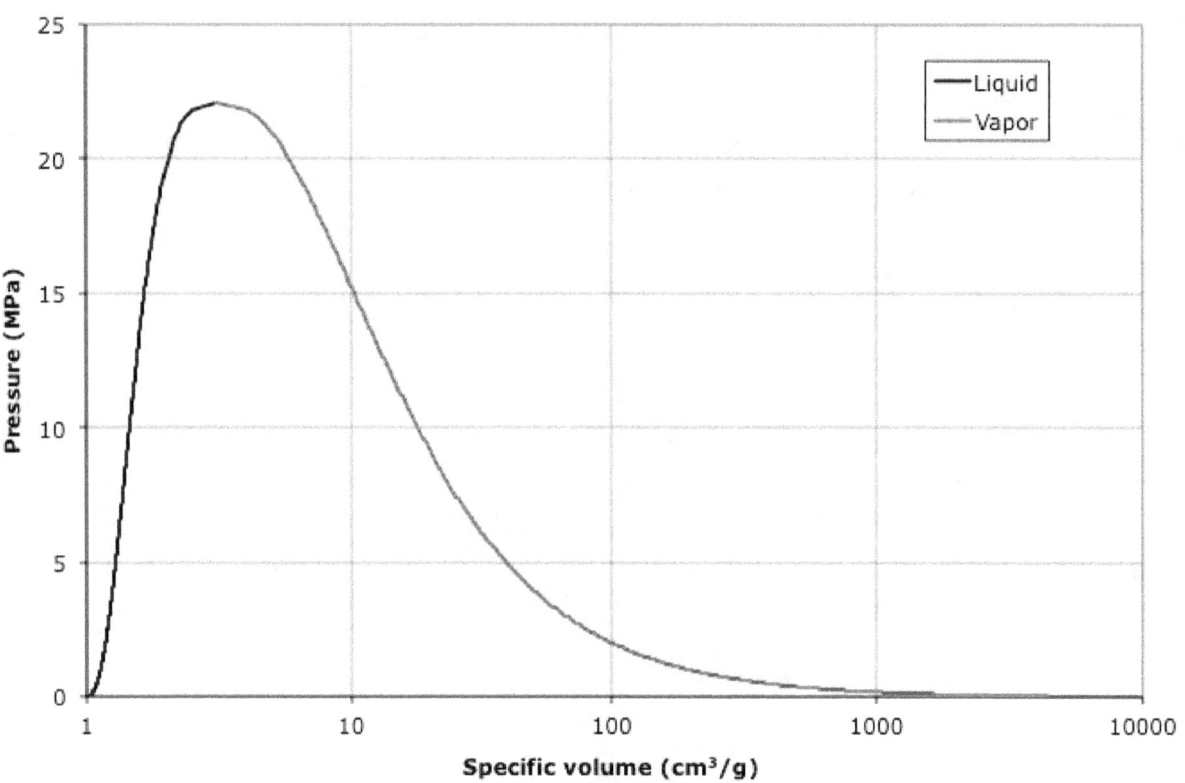

Figure A1: P-v vapor dome for Water (H_2O) on semi-log plot.

The following two figures show vapor domes plotted on T-S plots. Figure A2 shows water, a normal fluid with a mound-shaped vapor dome, while Figure A3 shows R134a, which is termed an isentropic fluid, because the right side of the vapor dome is nearly vertical (constant entropy). Isentropic fluids are often used as the working fluids in refrigerators and organic Rankine cycles. Why do you think that is?

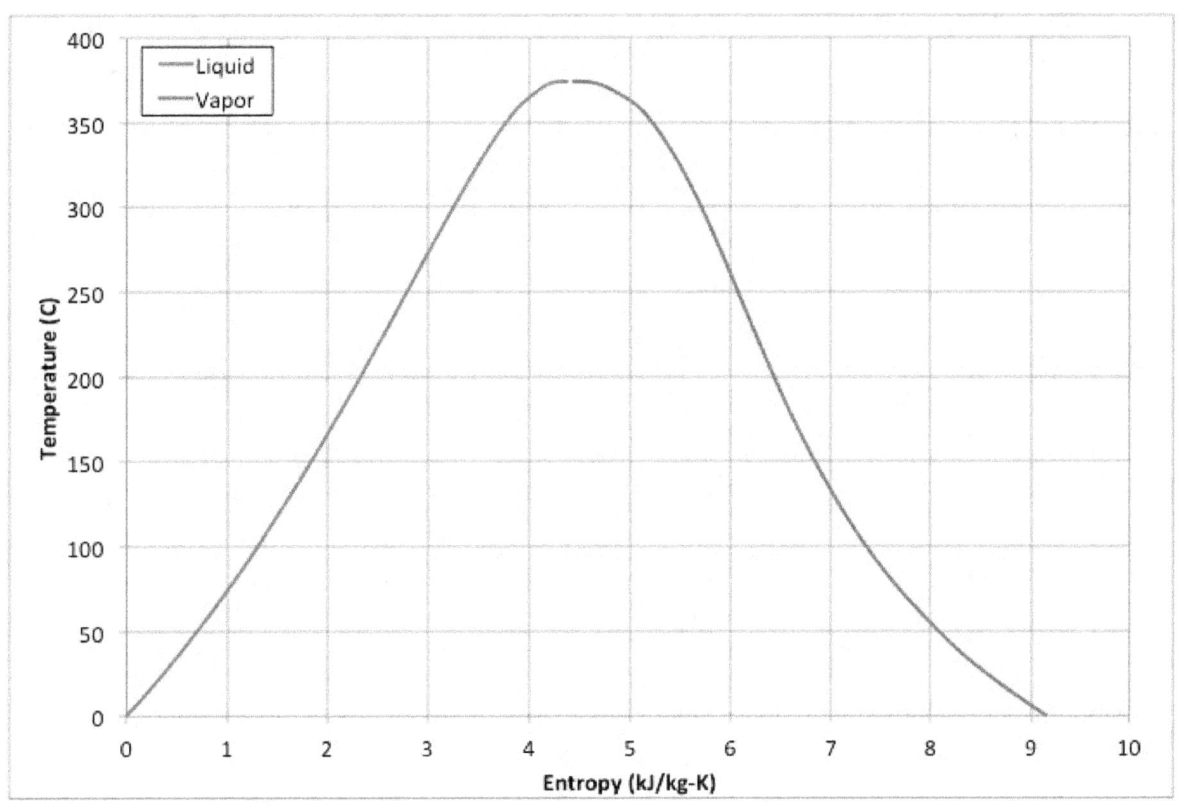

Figure A2: T-s vapor dome for water (H_2O).

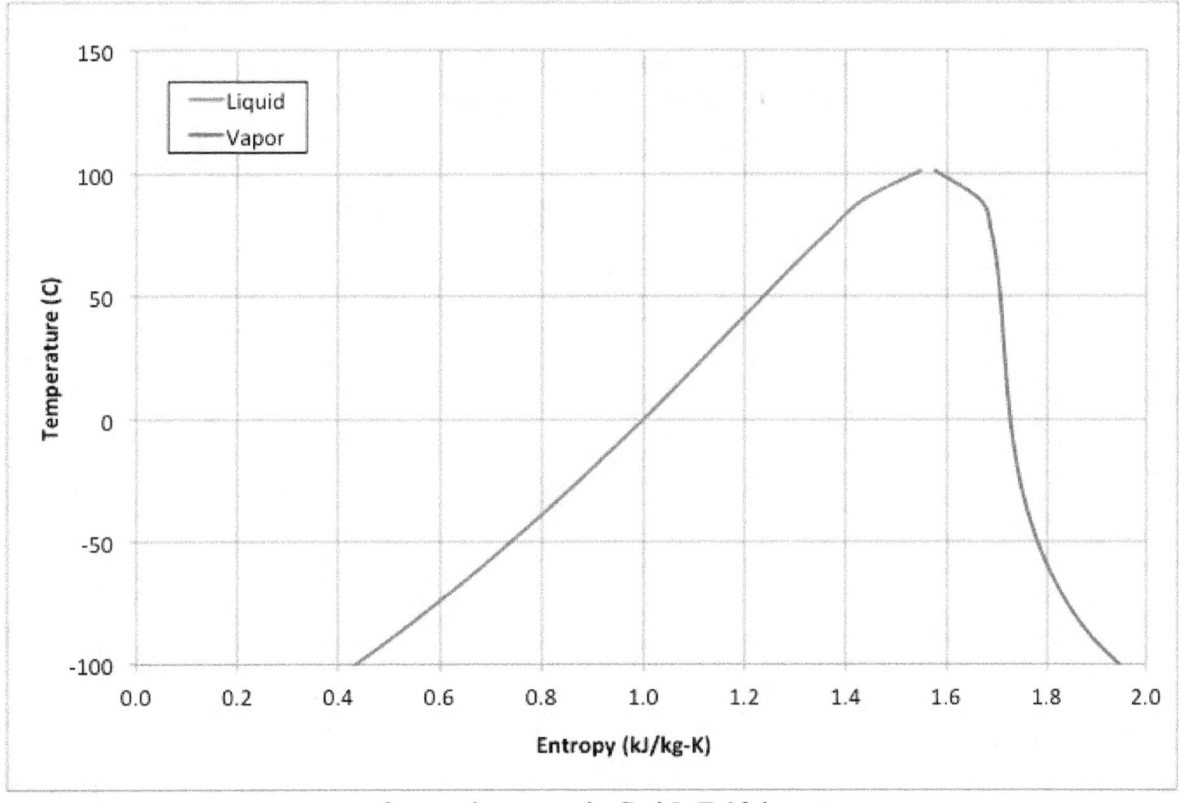

Figure A3: T-s vapor dome for an isentropic fluid, R134a.

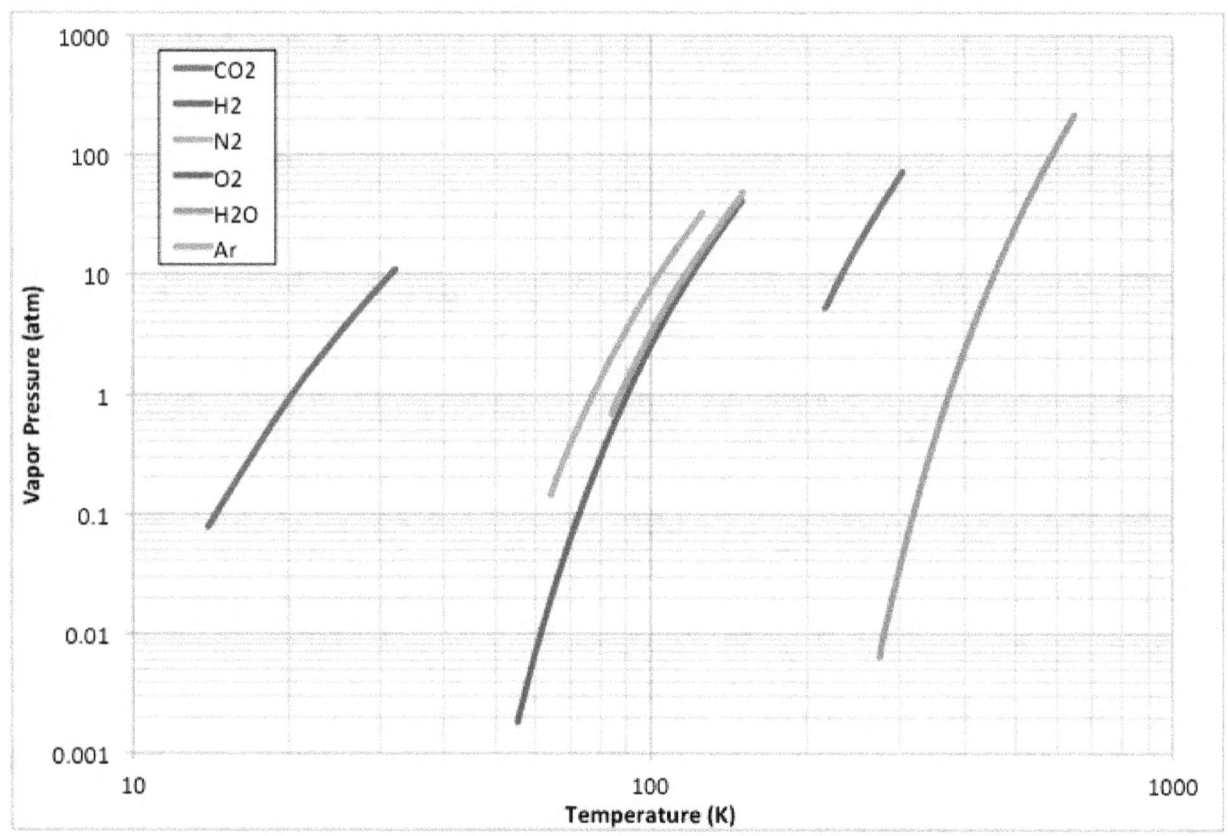

Figure A4: Vapor pressures for some simple substances on log-log plot.

118

Summary Tables
Table A1: Properties of common substances.

Name	Symbol	Molecular Weight (kg/kmol)	Triple Point T (K)	Normal Boiling Point (K)	Critical Temperature (K)	Critical Pressure (bar)	Critical density (kg/m^3)
Ammonia	NH_3	17.03	195.5	239.7	406.1	114.2	234.0
Argon	Ar	39.95	83.8	87.3	150.9	48.6	535.3
Carbon Dioxide	CO_2	44.01	216.6	194.8	304.1	73.8	467.4
Helium	He	4.00	2.2	4.2	5.2	2.3	69.6
Hydrogen	H_2	2.02	13.8	20.3	33.15	13.0	31.3
Methane	CH_4	16.04	90.7	111.7	190.6	46.0	162.6
Nitrogen	N_2	28.01	63.2	77.4	126.2	34.0	313.2
Oxygen	O_2	32.00	54.4	90.2	154.6	50.4	436.3
Propane	C_3H_8	44.10	85.0	231.1	369.8	42.5	220.0
Water	H_2O	18.01	273.2	372.8	647.1	220.6	322.0
Xenon	Xe	131.30	161.3	165.0	289.7	58.4	1102.9

Sources: Thermodynamic Properties of Minerals and Related Substances at 298.15 K and 1 Bar Pressure and at Higher Temperatures. R. Robie, B. Hemingway and J. Fisher. US Geological Survey Bulletin 1452. 1978. Thermodynamic Properties of Argon From the Triple Point to 300 K at Pressures to 1000 Atmospheres. NSRDS-NB 27. U.S. National Bureau of Standards. 1969. Tables of Thermodynamic Properties of Ammonia. Circular of the Bureau of Standards No. 142. 1923. Thermophysical Properties of Oxygen from the Freezing Liquid Line to 600 R for Pressures to 5000 Psia. NBS Technical Notes 384. 1971. NACA RM E56I21. Joule-Thomson Inversion Curves and Related Coefficients for Several Simple Fluids. R. Hendricks, I. Peller, A. Baron. NASA TN D-6807. 1972. National Bureau of Standards Circular 500, 1952

Table A2: Properties of selected combustion fuels.

Fuel	Formula	Molecular Weight (kg/kmol)	Stoic A/F	LHV (MJ/kg)	HHV (MJ/kg)
Hydrogen	H_2	2.02	34.3	120.0	142.2
Methane	CH_4	16.0	17.2	50.0	55.5
Propane	C_3H_8	44.1	15.7	46.3	50.2
Methanol	CH_3OH	32.0	6.5	20.1	22.9
Ethanol	C_2H_5OH	46.1	9.0	27.0	29.8
Gasoline	mixture	~102	14.7	42.4	45.4
Diesel Fuel	mixture	~200	14.7	42.6	45.6

Sources: Alternatives to Traditional Transportation Fuels. DOE/EIA-0585/0. 1994. NACA RM E56I21.
http://webbook.nist.gov

Properties of Water

Table A3: Density of Water Vapor as a function of temperature and pressure, in units of kg/m³.

T (K)	1 atm	10 atm	100 atm	200 atm	300 atm
400	0.555				
450	0.491				
500	0.441	4.597			
550	0.400	4.115			
600	0.367	3.739	50.96		
650	0.338	3.432	41.27	129.80	
700	0.314	3.175	36.06	88.38	190.3
750	0.293	2.955	32.43	73.60	129.8
800	0.275	2.765	29.66	64.72	107.3
850	0.258	2.598	27.46	58.46	93.94

Source: Tables of Thermal Properties of Gases. National Bureau of Standards Circular 564. 1955.

Table A4: Properties of compressed liquid water at 20 °C as a function of pressure.

P (bar)	Density (kg/m³)	Enthalpy (kJ/kg)	Entropy (kJ/kg-K)
0.1	998.17	83.92	0.29648
0.5	998.18	83.96	0.29647
1	998.21	84.01	0.29646
2	998.25	84.10	0.29644
4	998.34	84.29	0.29640
5	998.39	84.38	0.29638
6	998.44	84.48	0.29636
8	998.53	84.66	0.29632
10	998.62	84.85	0.29628
20	999.08	85.79	0.29607
40	999.99	87.67	0.29564
60	1000.89	89.54	0.29522
80	1001.80	91.41	0.29478
100	1002.69	93.28	0.29435
200	1007.13	102.57	0.29207
400	1015.75	120.90	0.28716
600	1024.02	138.94	0.28180
800	1031.97	156.70	0.27604
1000	1039.63	174.22	0.26992
2000	1074.00	258.73	0.23557
3000	1103.40	339.50	0.19783
4000	1129.10	417.72	0.15913
5000	1152.00	494.10	0.12060

Source: Harvey, A. Thermodynamic Properties of Water: Tabulation from the IAPWS Formulation 1995 for the Thermodynamic Properties of Ordinary Water Substance for General and Scientific Use. NISTIR 5078. 1998.

Table A5: Properties of saturated water as a function of temperature.

T °C	P bar	ρ_f kg/m³	ρ_g kg/m³	h_f kJ/kg	h_g kJ/kg	s_f kJ/kg-K	s_g kJ/kg-K
0.01	0.0061	999.79	0.00486	0	2500.9	0	9.1555
10	0.0123	999.65	0.00941	42.02	2519.2	0.1511	8.8998
20	0.0234	998.16	0.01731	83.91	2537.4	0.2965	8.6660
30	0.0425	995.61	0.03042	125.73	2555.5	0.4368	8.4520
40	0.0738	992.18	0.05124	167.53	2573.5	0.5724	8.2555
50	0.1235	988.00	0.08315	209.34	2591.3	0.7038	8.0748
60	0.1995	983.16	0.13043	251.18	2608.8	0.8313	7.9081
70	0.3120	977.73	0.19843	293.07	2626.1	0.9551	7.7540
80	0.4741	971.77	0.29367	335.01	2643.0	1.0756	7.6111
90	0.7018	965.30	0.42390	377.04	2659.5	1.1929	7.4781
100	1.0142	958.35	0.59817	419.17	2675.6	1.3072	7.3541
110	1.4338	950.95	0.82693	461.42	2691.1	1.4188	7.2381
120	1.9867	943.11	1.1221	503.81	2705.9	1.5279	7.1291
130	2.7028	934.83	1.4970	546.38	2720.1	1.6346	7.0264
140	3.6154	926.13	1.9667	589.16	2733.4	1.7392	6.9293
150	4.7616	917.01	2.5481	632.18	2745.9	1.8418	6.8371
160	6.1823	907.45	3.2596	675.47	2757.4	1.9426	6.7491
170	7.9219	897.45	4.1222	719.08	2767.9	2.0417	6.6650
180	10.028	887.00	5.1588	763.05	2777.2	2.1392	6.5840
190	12.552	876.08	6.3954	807.43	2785.3	2.2355	6.5059
200	15.549	864.66	7.8610	852.27	2792.0	2.3305	6.4302
210	19.077	852.72	9.5885	897.63	2797.3	2.4245	6.3563
220	23.196	840.22	11.615	943.58	2800.9	2.5177	6.2840
230	27.971	827.12	13.985	990.19	2802.9	2.6101	6.2128
240	33.469	813.37	16.749	1037.6	2803.0	2.7020	6.1423
250	39.762	798.89	19.967	1085.8	2800.9	2.7935	6.0721
260	46.923	783.63	23.712	1135.0	2796.6	2.8849	6.0016
270	55.030	767.46	28.073	1185.3	2789.7	2.9765	5.9304
280	64.166	750.28	33.165	1236.9	2779.9	3.0685	5.8579
290	74.418	731.91	39.132	1290.0	2766.7	3.1612	5.7834
300	85.879	712.14	46.168	1345.0	2749.6	3.2552	5.7059
310	98.651	690.67	54.541	1402.2	2727.9	3.3510	5.6244
320	112.84	667.09	64.638	1462.2	2700.6	3.4494	5.5372
330	128.58	640.77	77.050	1525.9	2666.0	3.5518	5.4422
340	146.01	610.67	92.759	1594.5	2621.8	3.6601	5.3356
350	165.29	574.71	113.61	1670.9	2563.6	3.7784	5.2110

360	186.66	527.59	143.90	1761.7	2481.5	3.9167	5.0536
370	210.44	451.43	201.84	1890.7	2334.5	4.1112	4.8012
374	220.64	322.00	322.00	2084.3	2084.3	4.4070	4.4070

Source: Harvey, A. Thermodynamic Properties of Water: NISTIR 5078. 1998.

Table A6: Properties of superheated water vapor as a function of temperature.

P = 0.1 bar

T °C	ρ kg/m^3	h kJ/kg	s kJ/kg-K
50	0.067263	2592.0	8.1741
60	0.065211	2611.2	8.2326
80	0.061474	2649.3	8.3439
100	0.058152	2687.5	8.4489
120	0.055176	2725.6	8.5484
140	0.052493	2763.9	8.6434
160	0.050060	2802.3	8.7341
180	0.047845	2840.8	8.8212
200	0.045818	2879.6	8.9049
240	0.042240	2957.8	9.0635
280	0.039182	3036.8	9.2118
320	0.036537	3116.9	9.3515
360	0.034228	3197.9	9.4837
400	0.032193	3279.9	9.6094
440	0.030386	3363.0	9.7293
480	0.028772	3447.2	9.8441
520	0.027320	3532.5	9.9544
560	0.026008	3618.8	10.061
600	0.024817	3706.3	10.163
640	0.023729	3794.9	10.262
680	0.022733	3884.6	10.358
720	0.021818	3975.5	10.452
760	0.020973	4067.5	10.543
800	0.020191	4160.6	10.631
840	0.019465	4254.9	10.717
880	0.018790	4350.2	10.802
920	0.018160	4446.7	10.884
960	0.017571	4544.2	10.964
1000	0.017019	4642.8	11.043
1200	0.014708	5150.7	11.413
1400	0.012950	5681.3	11.751
1600	0.011567	6231.1	12.061
1800	0.010452	6797.2	12.348
2000	0.009532	7377.0	12.615

P = 1.0 bar

T °C	ρ kg/m^3	h kJ/kg	s kJ/kg-K
100	0.58967	2675.8	7.3610
120	0.55767	2716.6	7.4678
140	0.52936	2756.7	7.5672
160	0.50402	2796.4	7.6610
180	0.48113	2836.0	7.7503
200	0.46031	2875.5	7.8356
240	0.42382	2954.6	7.9962
280	0.39280	3034.4	8.1459
320	0.36607	3114.9	8.2864
360	0.34278	3196.3	8.4191
400	0.32230	3278.6	8.5452
440	0.30414	3361.9	8.6653
480	0.28793	3446.2	8.7804
520	0.27337	3531.6	8.8908
560	0.26021	3618.0	8.9972
600	0.24827	3705.6	9.0998
640	0.23738	3794.3	9.1991
680	0.22740	3884.0	9.2954
720	0.21823	3975.0	9.3888
760	0.20977	4067.0	9.4797
800	0.20194	4160.2	9.5681
840	0.19468	4254.5	9.6544
880	0.18792	4349.9	9.7386
920	0.18162	4446.4	9.8209
960	0.17572	4543.9	9.9013
1000	0.17020	4642.6	9.9800
1200	0.14708	5150.6	10.350
1400	0.12950	5681.2	10.688
1600	0.11567	6231.0	10.998
1800	0.10451	6797.2	11.285
2000	0.09532	7377.0	11.552

Properties of superheated water vapor as a function of temperature (continued).

P = 10.0 bar

T °C	ρ kg/m³	h kJ/kg	s kJ/kg-K
180	5.1431	2777.4	6.5857
200	4.8539	2828.3	6.6955
240	4.3944	2920.9	6.8836
280	4.0322	3008.6	7.0482
320	3.7333	3094.4	7.1979
360	3.4801	3179.4	7.3367
400	3.2615	3264.5	7.4669
440	3.0704	3349.9	7.5902
480	2.9014	3435.8	7.7075
520	2.7507	3522.6	7.8196
560	2.6154	3610.1	7.9273
600	2.4931	3698.6	8.0310
640	2.3820	3788.0	8.1312
680	2.2805	3878.5	8.2281
720	2.1875	3970.0	8.3221
760	2.1018	4062.5	8.4135
800	2.0227	4156.1	8.5024
840	1.9494	4250.8	8.5890
880	1.8813	4346.5	8.6735
920	1.8178	4443.3	8.7560
960	1.7585	4541.1	8.8366
1000	1.7030	4639.9	8.9155
1200	1.4710	5148.9	9.2866
1400	1.2948	5680.0	9.6245
1600	1.1564	6230.3	9.9351
1800	1.0447	6796.7	10.222
2000	0.9528	7376.8	10.489

P = 100.0 bar

T °C	ρ kg/m³	h kJ/kg	s kJ/kg-K
320	51.894	2782.8	5.7133
360	42.873	2962.7	6.0075
400	37.827	3097.4	6.2141
440	34.312	3214.6	6.3833
480	31.619	3323.0	6.5311
520	29.441	3426.4	6.6649
560	27.619	3526.9	6.7886
600	26.057	3625.8	6.9045
640	24.694	3723.7	7.0142
680	23.490	3821.3	7.1187
720	22.414	3918.7	7.2189
760	21.444	4016.4	7.3153
800	20.564	4114.5	7.4085
840	19.760	4213.0	7.4986
880	19.022	4312.2	7.5861
920	18.342	4412.0	7.6712
960	17.712	4512.5	7.7541
1000	17.126	4613.8	7.8349
1200	14.719	5131.7	8.2126
1400	12.924	5668.7	8.5543
1600	11.528	6223.1	8.8671
1800	10.409	6792.4	9.1559
2000	9.490	7374.6	9.4239

Source: Harvey, A. Thermodynamic Properties of Water: Tabulation from the IAPWS Formulation 1995 for the Thermodynamic Properties of Ordinary Water Substance for General and Scientific Use. NISTIR 5078. 1998.

Properties of Air

Table A7: Density of Air as a function of temperature and pressure, in units of kg/m^3.

T (K)	0.01 atm	0.1 atm	1 atm	10 atm	100 atm
50	0.070688				
100	0.035305	0.35366	3.5985		
150	0.023533	0.23546	2.3673	25.0927	
200	0.017650	0.17654	1.7690	18.0715	217.748
250	0.014120	0.14121	1.4133	14.2506	149.928
300	0.011767	0.11767	1.1769	11.7990	118.455
350	0.010086	0.10086	1.0086	10.0818	99.112
400	0.008825	0.08825	0.8822	8.8069	85.690
450	0.007844	0.07844	0.7842	7.8203	75.707
500	0.007060	0.07060	0.7057	7.0354	67.923
600	0.005883	0.05883	0.5881	5.8601	56.519
700	0.005043	0.05043	0.5040	5.0235	48.502
800	0.004063	0.04412	0.4411	4.3963	42.514
900	0.003922	0.03922	0.3920	3.9089	37.873
1000	0.003530	0.03530	0.3529	3.5184	34.161
1250	0.002824	0.02824	0.2823	2.8162	27.460
1500	0.002353	0.02353	0.2353	2.3469	22.972
1750	0.002017	0.02017	0.2017	2.0133	19.747
2000	0.001762	0.01764	0.1764	1.7611	17.324
2250	0.001556	0.01564	0.1567	1.5659	15.427
2500		0.01397	0.1407	1.4081	13.904
3000		0.01100	0.1148	1.1650	11.591

Source: Tables of Thermal Properties of Gases. National Bureau of Standards Circular 564. 1955.

Table A8: Enthalpy of Air as a function of temperature and pressure, in units of kJ/kg.

T (K)	0.01 atm	0.1 atm	1 atm	10 atm	100 atm
50	49.76				
100	99.91	99.77	98.42		
150	150.02	149.95	149.26	142.01	
200	200.14	200.10	199.67	195.30	146.89
250	250.28	250.25	249.96	247.05	219.32
300	300.47	300.44	300.24	298.22	279.91
350	350.78	350.77	350.62	349.19	336.64
400	401.31	401.29	401.19	400.19	391.46
450	452.15	452.15	452.08	451.37	445.51
500	503.39	503.39	503.34	502.88	499.22
600	607.39	607.39	607.37	607.26	606.80
700	713.68	713.68	713.69	713.80	715.50
800	822.34	822.34	822.42	822.66	825.95
900	933.37	933.37	933.37	933.76	938.39
1000	1046.59	1046.59	1046.59	1047.06	1052.55
1250	1338.03	1338.03	1338.03	1338.66	1345.79
1500	1640.76	1640.45	1640.45	1641.23	1649.39
1750	1957.22	1954.55	1954.08	1954.63	1963.33
2000	2305.43	2286.53	2280.57	2279.47	2288.18
2250	2767.17	2661.40	2627.21	2617.96	2624.55
2500		3145.41	3016.35	2976.44	2973.78
3000			4108.81	3814.54	3727.28

Source: Tables of Thermal Properties of Gases. National Bureau of Standards Circular 564. 1955.

Table A9: Entropy of Air as a function of temperature and pressure, in units of kJ/kg-K.

T (K)	0.01 atm	0.1 atm	1 atm	10 atm	100 atm
50	6.3903				
100	7.0857	6.4236	5.7541	5.1046	
150	7.4916	6.8303	6.1665	5.4728	
200	7.7800	7.1187	6.4566	5.7799	4.9318
250	8.0036	7.3426	6.6808	6.0109	5.2590
300	8.1867	7.5257	6.8642	6.1983	5.4803
350	8.3417	7.6810	7.0197	6.3545	5.6550
400	8.4765	7.8159	7.1546	6.4908	5.8014
450	8.5965	7.9356	7.2746	6.6113	5.9288
500	8.7044	8.0435	7.3825	6.7198	6.0431
600	8.8938	8.2329	7.5719	6.9101	6.2382
700	9.0577	8.3968	7.7358	7.0743	6.4058
800	9.2027	8.5417	7.8810	7.2195	6.5531
900	9.3335	8.6726	8.0116	7.3501	6.6851
1000	9.4526	8.7917	8.1307	7.4692	6.8045
1250	9.7127	9.0517	8.3907	7.7292	7.0645
1500	9.9331	9.2721	8.6114	7.9493	7.2841
1750	10.1277	9.4658	8.8046	8.1422	7.4778
2000	10.3131	9.6432	8.9785	8.3155	7.6508
2250	10.5300	9.8194	9.1415	8.4748	7.8090
2500		10.0226	9.3054	8.6258	7.9559
3000			9.7003	8.9297	8.2306

Source: Tables of Thermal Properties of Gases. National Bureau of Standards Circular 564. 1955.

From perusing these data tables in the Appendices, you may notice a few things:
- From the density tables you can notice that air very much behaves as an ideal gas (increase the pressure by a factor of 10, and the density should increase by 10, at fixed temperature)
- Steam (H_2O) does not behave as an ideal gas
- Gas behavior is closer to ideal gas law at higher temperatures and lower pressures
- Entropy increases with increasing temperature, but decreases with increasing pressure
- Enthalpy depends more strongly on temperature than pressure (for a truly Ideal Gas, it does not depend on pressure at all)

Conversion Factors

English to Metric	Metric to English
1 lbm = 0.4536 kg	1 kg = 2.205 lbm = 0.0873 slug
1 ft = 12 in = 0.3048 m = 30.48 cm	1 m = 3.28 ft
1 Gallon = 231 in^3 = 3.76 L	1 m^3 = 1000 L = 35.3 ft^3
1 mi/hr = 1.609 km/hr = 0.447 m/s	1 m/s = 3.28 ft/s = 2.237 mph
1 ft^3/s = 0.0283 m^3/s	1 L/min = 0.264 Gallon/minute
1 lbf = 4.45 N	1 N = 0.225 lbf
1 psi = 6895 Pa	1 Pa = 0.000145 psi
1 ft-lbf = 1.357 N-m	1 N-m = 0.737 ft-lbf
1 Btu = 778 ft-lbf = 1055 J	1 J = 0.000948 Btu
1 hp = 550 ft-lbf/s = 746 W	1 kW = 1.341 hp
[°F] = 9/5 [°C] + 32	[°C] = ([°F] - 32)*5/9

Conversion factors: 1 hp = 550 ft-lbf/s   4.45 N = 1 lbf   1 m = 3.28 ft
1 lbf = 32.2 lbm-ft/s^2 1 hp = 746 W   1 m/s = 2.24 mph   1 mile = 5280 ft
1 Gal = 3.78 L   1000 L = 1 m^3 2.54 cm = 1 in   1 kg = 2.2 lbm

Other Commonly Used Quantities

g = 9.8 m/s^2 = 32.2 ft/s^2.

R = 8314 J/kmol-K = 10.73 ft^3-psi/lbmol-°R = 1545 ft-lbf/ lbmol-°R

1 lbf = 32.2 lbm-ft/s^2 = 1 slug-ft/s^2

1 N = 1 kg-m/s^2

MATLAB CODE

```
clear all
clc
% this code uses IAPWS formulation for properties of saturated water
% properties of water
Tc = 647.1;        % K
Pc = 22064000;     % Pa
rhoc = 322;        % kg/m3
a0 = 1;            % kJ/kg
% specify temperature in K
T = 373.15
if (T > Tc)
    stop
end
q = T/Tc;
t = 1 - T/Tc;
% constants
a1 = -7.85951783;
a2 = 1.84408259;
a3 = -11.7866497;
a4 = 22.6807411;
a5 = -15.9618719;
a6 = 1.80122502;
b1 = 1.99274064;
b2 =  1.09965342;
b3 = -0.510839303;
b4 = -1.75493479;
b5 = -45.5170352;
b6 = -674694.450;
c1 = -2.03150240;
c2 = -2.68302940;
c3 = -5.38626492;
c4 = -17.2991605;
c5 = -44.7586581;
c6 = -63.9201063;
d1 = -0.0000000565134998;
d2 = 2690.66631;
d3 = 127.287297;
d4 = -135.003439;
d5 = 0.981825814;
```

```
da = -1135.905627715;
dp = 2319.5246;
% vapor pressure
Pvap=Pc*exp((Tc/T)*(a1*t+a2*t^1.5+a3*t^3+a4*t^3.5+a5*t^4+a6*t^7.5))
% dP/dT
dpdt=-Pvap/T*(a1+1.5*a2*t^0.5+3*a3*t^2+3.5*a4*t^2.5+4*a5*t^3+7.5*a6*t^6.5+log(Pvap/Pc));
% density of saturated liquid kg/m3
rhof=rhoc*(1+b1*t^(1/3)+b2*t^(2/3)+b3*t^(5/3)+b4*t^(16/3)+b5*t^(43/3)+b6*t^(110/3))
% density of saturated vapor  kg/m3
rhog=rhoc*exp(c1*t^(1/3)+c2*t^(2/3)+c3*t^(4/3)+c4*t^3+c5*t^(37/6)+c6*t^(71/6))
% specific volume of saturated vapor m3/kg
vg = 1/rhog
% enthalpy of saturated liquid kJ/kg
hf=(da+d1*q^-19+d2*q+d3*q^4.5+d4*q^5+d5*q^54.5)+T/rhof*dpdt/1000
% enthalpy of saturated liquid kJ/kg
hg=(da+d1*q^-19+d2*q+d3*q^4.5+d4*q^5+d5*q^54.5)+T/rhog*dpdt/1000
% entropy of saturated liquid
sf=(dp+19/20*d1*q^-20+d2*log(q)+9/7*d3*q^3.5+5/4*d4*q^4+109/107*d5*q^53.5)/Tc+dpdt/rhof/1000
% entropy of saturated vapor
sg=(dp+19/20*d1*q^-20+d2*log(q)+9/7*d3*q^3.5+5/4*d4*q^4+109/107*d5*q^53.5)/Tc+dpdt/rhog/1000
```